# OIL

## Final Countdown
## To A Global Crisis

▼

## And Its Solutions

By
## Dr. Sahadeva dasa

B.com., FCA., AICWA., PhD
Chartered Accountant

SOUL
SCIENCE UNIVERSITY

Soul Science University Press
OilcrisisSolutions.com

A Vedic Prospective on Energy Crisis, Hydrocarbon Dependence and
Alternative Living

Readers interested in the subject matter of this
book are invited to correspond with the publisher at:
SoulScienceUniversity@gmail.com   +91 98490 95990


To order a copy write to chandra@rgbooks.co.in
or buy online: at www.rgbooks.co.in


First Edition: June 2008


Soul Science University Press expresses its gratitude to the Bhaktivedanta
Book Trust International (BBT), for the use of quotes by His Divine Grace
A.C.Bhaktivedanta Swami Prabhupada.
Copyright Bhaktivedanta Book Trust International (BBT)

ISBN  978-81-909760-0-8


Published by:

Dr. Sahadeva dasa for Soul Science University Press


Designed by : Sailesh Ijmulwar,  Waar Creatives.

Printed in India : Rainbow Print Pack, Hyderabad

*Dedicated to*

His Divine Grace A.C. Bhaktivedanta Swami Prabhupada

*"That is the way of material civilization, too much depending on machine. At any time the whole thing may collapse and therefore we may not be self complacent depending so much on artificial life. The modern life of civilization depends wholly on electricity and petrol and both of them are artificial for man."*
*~Srila Prabhupada (Letter, 13 November, 1965)*

By The Same Author

*End of Modern Civilization and Alternative Future*

*To Kill Cows Means To End Human Civilization*

*Cow And Humanity – Made For Each Other*

*Capitalism Communism And Cowism – A New Economics*
*For The 21st Century*

*Cows Are Cool – Love 'Em !*

*Wondrous Glories of Vraja*

*Modern Foods – Stealing Years From Your Life*

*Noble Cow – Munching Grass, Looking Curious And Just Hanging Around*
*Lets Be Friends – A Curious, Calm Cow*
*We Feel – Just Like You Do*
(More information on availability at the back)

# Contents

*Preface*

Industrial civilization which began 150 years ago is entering a decisive phase as the resources run scarce to satisfy its ravenous appetite. Human society, living an industrialized life, has used up resources which took nature millions of years to create. How long our earth and environment can support this reckless living is a big question now.

Human beings constitute a very insignificant portion of the cosmos and acting up like masters of the universe is an illusion. Thinking that we can fight nature is a hallucination. Living a life in harmony with nature and its environment is the only way a civilization can survive long enough to pass the test of time.

On many fronts, the crumbling of this colossal industrial setup is becoming apparent. One on the forefront is peak oil or upcoming oil shortage. Oil is the lifeblood of modern civilization. Choke off the oil and it quickly seizes.

With no viable alternatives in sight, human society will be facing a great crisis of unprecedented scale. All the previous calamities were local in nature. Oil crisis would be a global disaster because the world today shares a common fate, thanks to interdependence and interconnectivity. Earlier we suffered in isolation and now we go down, all together. There are food riots in Africa when America decides to produce bio-fuels. When there are skirmishes in Nigeria, a government is toppled on other side of the globe.

The solution lies in aligning ourselves with nature rather than trying to put up a brave front. So called victory over nature is just a garbed defeat. We have to recognize the problem as overconsumption and not overpopulation.

This book is an awakening call, a call to act before time runs out, before it's too late. The saying "man is the architect of his own fortune" was never more relevant.

President Bush aptly admits and endorses this view when he describes the present times for America and rest of the world as "very difficult times, very difficult."

The headlines have begun to blare like "Shell chief fears oil shortage in seven years", "Spectre of food rationing hits US amidst global food crisis ", "Futures Market Traders Bet On $200/Barrel Oil In 2008", "Oilcos plan to ration fuel supply", "British truckers protest rising fuel prices", " Oil shock: Airlines cut flights, expansion plans", etc.

Stopgap solutions can not help as the nature and environment can not be cheated. In an attempt to do so, we end up cheating our ownselves. We have to get to the root of the problem which is an unsustainable lifestyle. Present day lifestyle is founded on total disregard for the nature and its Creator.

*Sahadeva dasa*

Dr. Sahadeva dasa

June 25, 2008
Secunderabad, India

# Section-I

Oil - Introduction

# What You Live On Today Is Oil

*'Well, the secret of success is to get up early, work late – and strike oil.'*
*–Rockefeller*

Although we can trace the beginnings of oil itself to several million years ago, the oil industry is a comparatively recent development.

Petroleum literally means 'rock oil'. It is the second most abundant liquid on Earth. Oil and gas also provide two-thirds of the world's primary energy supplies. Oil and gas are also non-renewable resources and our use of them has increased so much that we have worries about how long they will last.

First coal and now petroleum (which includes oil and gas) have played an essential role in changing our society from an agricultural to an industrial one. It is almost impossible to find any synthetic item where petroleum has not had any part in the process of its manufacture.

From the moment we wake in the morning to the moment we go to sleep, oil controls our lives. Its influence reaches far into politics, international affairs, global economics, human rights and the environmental health of our planet.

Burning crude oil itself is of limited use. To extract the maximum value from crude, it first needs to be refined into petroleum products. The best-known of these is gasoline, or petrol. However, there are many other products that can be obtained when a barrel of crude oil is refined. These include liquefied petroleum gas (LPG), naphtha, kerosene, gasoil and fuel oil. Other useful products which are not fuels can also be manufactured by refining crude oil, such as lubricants and asphalt (used in paving roads). A range of sub-items like perfumes and insecticides are also ultimately derived from crude oil.

Furthermore, several of the products listed above which are derived from crude oil, such as naphtha, gasoil, LPG and ethane, can themselves be used as inputs or feedstocks in the production of petrochemicals. There are more than 4,000 different petrochemical products, but those which are considered as basic products include ethylene, propylene, butadiene, benzene, ammonia and methanol. The main groups of petrochemical end-products are plastics, synthetic fibres, synthetic rubbers, detergents and chemical fertilisers.

Considering the vast number of products that are derived from it, crude oil is a very versatile substance. Life as we know it today would be extremely difficult without crude oil and its by-products.

The most obvious way that oil dominates us, of course, is transportation. Oil powers 90 percent of world's transportation needs. Biggest consumer of oil is United States of America. It has more automobiles than any other country; in fact, it has more cars and trucks than it has people.

Besides, oil provides heat in the winter for millions of homes, we depend on oil for food, pharmaceuticals, chemicals and the entire bedrock of modern life. Without oil there would be no plastics, nor many of the chemical-based medicines we take for granted. Perhaps most important, we would go hungry without oil: commercial agriculture would grind to a halt without oil to run farm and food processing machinery or to make fertilizers, herbicides and pesticides.

We know that petroleum is drawn from deep wells and distilled

*And what about the next generation? If there is no petrol, what will the next generation do?*
*-Srila Prabhupada (Morning Walk, 29 April, 1973)*

into gasoline, jet fuel, and countless other products that form the lifeblood of industrial civilization and the adrenaline of military might.

A quick look at some facts on the importance of Petroleum, rightly called 'the black gold'.

-A reduction of as little as 10 to 15 percent in supply of petroleum can cripple oil-dependent industrial economies. In the 1970s, a reduction of just 5 percent caused a price increase of more than 400 percent.

*Crude oil is source of more than 4000 different petrochemical products.*

-Most farming equipment are either built in oil-powered plants or use diesel as fuel. Nearly all pesticides and many fertilisers are made from oil.

-Most plastics, used in everything from computers and mobile phones to pipelines, clothing and carpets, are made from oil-based substances.

-Manufacturing requires huge amounts of fossil fuels. The construction of a single car in the US requires, on an average, at least 20 barrels of oil.

-Most renewable energy equipment require large amounts of oil to produce.

-Metal production, particularly aluminium, cosmetics, hair dye, ink and many common painkillers all rely on oil.

-Oil allowed for the mass production of pharmaceutical drugs, and the development of health care infrastructure such as hospitals, ambulances, roads, etc.

-It is also required for nearly every consumer item, water supply pumping, sewage disposal, garbage disposal, street/park maintenance, hospitals and health systems, police, fire services and national defence.

*"Petroleum's worth as a portable, dense energy source powering the vast majority of vehicles and as the base of many industrial chemicals makes it one of the world's most important commodities. Access to it was a major factor in several military conflicts in last one hundred years."*
*~Joseph Moore*

## Check Out
## What's Flowing In Your Veins
## It's Oil

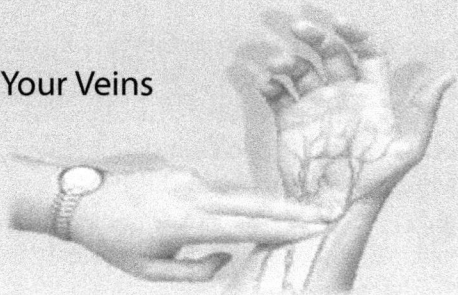

We have allowed oil to become vital to virtually everything we do. Ninety-five per cent of all goods in shops involve the use of oil. Ninety-five per cent of all our food products require oil use.

The three main purposes for which oil is used worldwide are food, transport and heating. In the near future the competition for oil for these three activities will be raw and real. Reliable supplies of cheap oil and natural gas underlie everything we identify as the necessities of modern life - not to mention all of its comforts and luxuries: central heating, air conditioning, cars, airplanes, electric lights, inexpensive clothing, recorded music, movies, hip-replacement surgery, national defense - you name it.

Imagine a day of your life without oil! Well its almost impossible to escape oil's influence for even a single day in our lives. Did you wake up to a plastic alarm clock after a restful night? Thank petroleum for the mosquito repellent, both bottle and liquid. And did you put on your eye glasses? Then you started your this day with petroleum too. The frame and plastic lens owe their origin to oil. When you get ready to shave, your shaving cream, razor body and deodorant contain petroleum products as important ingredients. Your bathroom pvc door and toilet seats, where do they come from? You guessed it right. Then comes toothbrush, an outright petroleum based product and toothpaste, with petrochemical-enhanced artificial coloring and mineral oils. You are living with oil in your mouth if you are wearing dentures as they are mostly petroleum based. After shower when you put your lip balm, you have used a petroleum product once again.

While still in shower, you rush to answer a phone call and it is all oil based plastic. Suddenly your tiny tot requires a change of diapers and its linings are fathered by petroleum. After shower, when you put on your formals, you are again draped in oil because all synthetic fabrics originate from oil. When you put on your leather shoes with synthetic soles, once again you step into the realms of oil . Then you quickly spray perfume. Of course, its oil. Then to avoid drizzle, you put on your raincoat and lo and behold! You have added another layer of oil on your existence.

But what about streets. You guessed it right! Streets are paved with asphalt, a sticky by-product remaining after refining crude oil. Of course no need to discuss what goes in your car to run and lubricate it. In your office canteen, your breakfast comes off a non stick pan and this too is a petroproduct. Of course food production and transport is also at the mercy of petroleum because most of the fertilizers and pesticides are harvested from oil. Then as you insert a CD or DVD, you have handled oil once again. So is with your credit and debit cards, bunch of which your wallet holds. And yet same again with your wallet whether it is leather or rexine. Leather too requires petrochemicals for tanning and processing. Well if all this is beginning to give you a headache and you would like to pop in an aspirin, guess where it comes from. Answer again would be the same.

Petroleum follows you when in the evening you head for a round of golf. Golf balls are practically oil solidified. We are merged or shall we say drowning, fro m toe to head, in an ocean of oil. Unfortunately this ocean is limited in its dimensions and full of fearsome waves of uncertain supplies.

Thus almost every current human endeavour from transportation, to manufacturing, to electricity to plastics, and especially food production is inextricably intertwined with oil and natural gas supplies.

*"Oil is like a gun, one's perception of it is very much dependent on which end of the barrel one finds oneself."*
*~Juan G. Carbonel*

Let us look at the world of transport and its tryst with oil. Following are a few examples from a list which is practically interminable:

| | |
|---|---|
| Brake | Jet Fuel |
| Fluid Boats | Diesel |
| Oil Filters | Tyres |
| Car battery cases | Paints |
| Car Bodies | Wire Coating |
| Motorcycle helmets | Asphalt Highways |
| Gasoline | Seats |

You can clearly perceive that our involvement with oil is deeper than most of us perceive. Now let us have a look at another area of household items.

| | |
|---|---|
| Candles | Curtains |
| Shower Doors | House Paint |
| Roofing | Fan Belts |
| Vinyl Wallpaper & Floorings | Telephones |
| Trash Bags | Detergents |
| Ice Cube Trays | Bowls |
| Blankets (synthetic) | Glue |
| Toilet Seats | Flooring |
| VCR & CassetteTapes | Carpets |
| Shampoo | Sofa and Chair Fittings. |

TODAY'S LECTURE 'MAN, MASTER OF THE UNIVERSE' HAS BEEN CANCELLED BECAUSE OF THE NATURAL GAS SHORTAGE.

Well, if still in doubt, do look at another list of day to day items, which are hand in gloves with petroleum industry.

| | |
|---|---|
| Toothbrush | Medicines |
| Deodorant | Containers |
| Eye Glasses | Clothesline |
| Nail Polish | Fertilizer |
| Hair Spray | Insect Repellent |
| Sweaters | Artificial Turf |
| Shoes | Herbicides & Insecticides |
| Sunglasses | Awnings |
| Crayons | Containers |
| Pajamas | GardenHose |
| Pillows | PVC pipes |
| Combs | Glue |
| Lipstick | Umbrella |
| Shampoo | |

*The simple fact is we cannot plant dead dinosaurs underneath our continental United States to create oil. It is simply not there.*
*~ Jay Inslee*

9

A green planet is today blackened by oil which has seeped into every nook and crevice of this planet. No region is free from oil dependence.

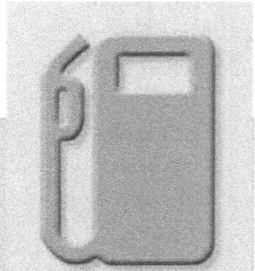

*U.S. gasoline consumption of 320,500,000 gallons per day (March 2005) worked out to about 3700 gallons or 14000 litres per second. If lined up in 1-gallon cans, they would encircle the earth at the equator almost 6 times (about 147,000 miles of cans) — every day. Here's another image: EVERY DAY, the US consumes enough oil to cover a football field with a column of oil 2500 feet tall. World oil demand has reached 85.4 million barrels a day in 2007. This simply means that the world today uses 157000 litres of oil per second.*

## Top Oil Consuming Nations

2007

| Rank | Countries | Amount (Barrels Per day) |
|------|-----------|--------------------------|
| 1 | United States: | 20,730,000 bbl/day |
| 2 | China: | 6,534,000 bbl/day |
| 3 | Japan: | 5,578,000 bbl/day |
| 4 | Germany: | 2,650,000 bbl/day |
| 5 | Russia: | 2,500,000 bbl/day |
| 6 | India: | 2,450,000 bbl/day |
| 7 | Canada: | 2,294,000 bbl/day |
| 8 | Korea, South: | 2,149,000 bbl/day |
| 9 | Brazil: | 2,100,000 bbl/day |
| 10 | France: | 1,970,000 bbl/day |
| 11 | Mexico: | 1,970,000 bbl/day |
| 12 | Italy: | 1,881,000 bbl/day |
| 13 | Saudi Arabia: | 1,845,000 bbl/day |
| 14 | United Kingdom: | 1,827,000 bbl/day |
| 15 | Spain: | 1,573,000 bbl/day |
| 16 | Iran: | 1,510,000 bbl/day |
| 17 | Indonesia: | 1,168,000 bbl/day |

One barrel equals to 159.98 liters. When oil was first excavated in US by one Mr Drake, it was collected in whiskey barrels and the 42 gallon measure has remained the industry standard ever since.

*"The oil can is mightier than the sword."*
*~ Everett Dirksen*

"Just like people are struggling. Wherever you go, material world, either you go to London or go to Paris or to Calcutta or Bombay, anywhere you go, what is the business? Everyone is struggling: (makes sounds) whoon, shoon, shoon, shoon, shoon, shoon, shoon. Day and night the motorcar going this way and that way, this way and that way. Wherever, we see this nonsense thing, whoo, shoo, shoon, shoon, shoon, shoon, shoo, shoo, shoo. Any city you go, the same road, same motorcar, same "whoo, shoosh," same petrol, that's all. What is the difference? But we say—this is called illusion—"I have come to Paris. I have come to Calcutta." But where is the difference between Calcutta and Paris and Bombay? The same thing. Punah punah carvita-carvananam. Again and again, chewing the chewed."
~Srila Prabhupada ((Lecture on Bhagavad-gita, Paris, August 13, 1973)

12

# Section-II

# Oil

# Setting The Scene For Wars

## In The Past, In The Present And In The Future

# Oil - A Historical Perspective

## Oil In Antiquity

Petroleum, in some form or other, is not a substance new in the world's history. More than four thousand years ago, according to Herodotus and confirmed by Diodorus Siculus, asphalt was employed in the construction of the walls and towers of Babylon. There were oil pits near Ardericca (near Babylon) and a pitch spring on Zacynthus. Great quantities of it were found on the banks of the river Issus, one of the tributaries of the Euphrates. Ancient Persian tablets indicate the medicinal and lighting uses of petroleum in the upper levels of their society.

The first oil wells were drilled in China in the 4th century or earlier. They had depths of up to 243 meters (about 800 feet) and were drilled using bits attached to bamboo poles. The oil was burned to evaporate brine and produce salt. By the 10th century, extensive bamboo pipelines connected oil wells with salt springs. The ancient records of China and Japan are said to contain many allusions to the use of natural gas for lighting and heating. Petroleum was known as burning water in Japan in the 7th century.

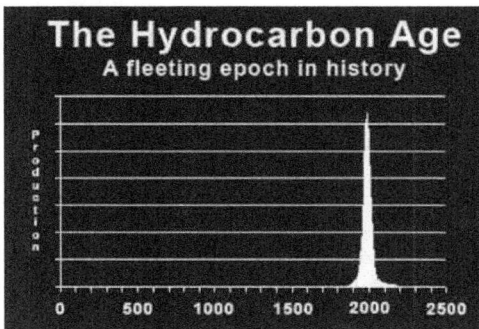

**The Hydrocarbon Age**
A fleeting epoch in history

In the 8th century, the streets of the newly constructed Baghdad were paved with tar, derived from easily accessible petroleum from natural fields in the region. In the 9th century, oil fields were exploited in the area around modern Baku, Azerbaijan, to produce naphtha. These fields were described by the geographer Masudi in the 10th century, and by Marco Polo in the 13th century, who described the output of those wells as hundreds of shiploads. Petroleum was first distilled by chemists in the 9th century, producing chemicals such as kerosene.

The earliest mention of American petroleum occurs in Sir Walter Raleigh's account of the Trinidad Pitch Lake in 1595; whilst thirty-seven years later, the account of a visit of a Franciscan, Joseph de la Roche d'Allion, to the oil springs of New York was published in Sagard's Histoire du Canada. A Russian traveller, Peter Kalm, in his work on America published in 1748 showed on a map the oil springs of Pennsylvania.

The modern history of petroleum began in 1846 with the discovery of the process of refining kerosene from coal by Atlantic Canada's Abraham Pineo Gesner.

The petroleum era in America was heralded by Edwin Drake's drilling of a shallow oil well in Titusville 1859.

## Modern History of Petroleum
### An American Beginning

In 1857, a small group investors formed the Pennsylvania Rock Oil Company to drill a liquid flammable substance that seeped from the earth around the hills in certain parts of Pennsylvania. The investors had valid interests. Household fuel supplies were dwindling due to large scale killing of whales.

The crude technique employed by the Pennsylvania Rock Oil Company to extract first crude was the same that had been used in salt wells in America. When the 'board' was staying in a hotel, they spotted a fellow called Edwin Drake who posed himself as one of the go-getter types. The company zeored in on Drake to go ahead with drilling in Pennsylvania, a task that would change United States and the world and alter the course of history.

*I mean, Iraqis feel that if we drove smaller cars, their lives would be longer. ~ Bill Maher*

To make Drake appear more than he really was, the company referred to him as 'Colonel Drake' in correspondence and he came to be known as such. In 1858, he started drilling near a place called Titusville. After a frustrating six months search and after having all the money blown up, despondent investors mailed him to shut the shop down. But the letter arrived late and in the meantime the 'Colonel' tried his luck for one last time and there appeared this seamless flow of oil. This marked the birth of oil industry and an industrial civilization based thereupon.

## Inception of A Corporatised World

Oil is responsible for giving birth to companies which would, in times to come, grow bigger than many economies of the world. This beginning was heralded by surfacing of a personality called John D. Rockefeller of Ohio. Rockefeller saw great fortunes in distribution of petrol rather than in just production.

Year 1870 marked the birth of Standard Oil Company with Mr Rockefeller at the top and it was not long before the company was dictating the fate of US oil industry. By building a strong cash position, the company manipulated the markets without much scruples and evolved into what we call a multinational today. Its manipulative practices drove many competitors to bankruptcy.

Three decades of Standard Company's reign finally woke the press up and it launched a scathing attack resulting in the US government's massive lawsuit against the company. A great legal battle ensued which resulted in declaration of the company as monopoly and its splitting in 34 companies. But in course of time, these split subsidiaries too grew in monstrous proportions. Global oil giants of today like Exxon, Chevron and Mobil are none other than these Standard's offsprings known formerly as Standard Oil of New Jersey, Standard Oil of New York and Standard Oil of California etc.

By the end of the century, companies such as General Motors, Wal-Mart, Exxon Mobil, Ford, and Daimler Chrysler were richer than entire nations. By the year 2000, of the world's one hundred leading economies, fifty-one were corporations.

## Europe Catches On

It did not take too long for Europe to join the oil club. In the last quarter of 19th century, Europe was utilizing US kerosene and it was a hefty source of revenue for US, more so for Standard Oil. But with the dawn of 20th century, usage of oil greatly increased in Europe and this attracted several players. Next was the find of oil in Baku, Russia. Rothschild of France headed for Asia after his attempts to takeover European markets were frustrated by Standard Oil. In the mean time, one Mr Marcus Samuel founded 'Shell' in London. Then there was 'Royal Dutch". Thus began the tussle for corporatization of the world.

Of course oil did not confine itself to domestic users. It was not too long before the oil trickled down in the military brains. Britain, Germany and several others quickly realized the superiority of oil over coal in powering battle fleets. None of these powers had oil of their own and still the advantages petroleum were too numerous to ignore. Of course, in times to come, this inevitably would give rise to a branch of diplomacy known as oilomacy. These great powers would do anything conceivable to retain their hold on world's oil fields.

Soon oil became the military life line. Petroleum made a hell and heaven difference to military paraphernalia. Speed, acceleration, light weight, cheapness, all these qualities of this wonder liquid went on to introduce a new concept of war to humanity, a war which was far more lethal, a war which could quickly engulf the globe as a whole.

Availability of petroleum ushered in an unprecedented race amongst two great powers of the time, Germany and Britain.

Wilhelm II of Germany desired to construct a formidable German navy which could tie in with German ambitions in the colonial and commercial spheres, threatening British domination in these areas. The Kaiser entrusted the establishment of this German navy to his Naval Minister and close advisor, Grand Admiral Alfred von Tirpitz.

Motivated by Wilhelm's backing and his own enthusiasm for an

*"Energy matters."*
*- Einstein 1905*

expanded navy, Tirpitz championed four Fleet Acts from 1898 to 1912. The German program was enough to alarm the British and drive them into the alliances with France and Russia.

Under the direction of Admiral Jackie Fisher, the First Sea Lord from 1903 to 1910, the Royal Navy embarked on its own massive expansion to keep ahead of the Germans. The cornerstone of British naval rearmament was to be the revolutionary battleship Dreadnought, which was launched in 1906. From then on until 1914, the British and Germans vied with each other to construct superior numbers of battleships, submarines, and other naval vessels and weaponry. This arms race, based on imperialistic ambitions just needed a spark to explode into a great ruinous war.

## First World War

### First Oil Propelled War

Modern warfare, which inflicts far more casualties than a conventional one, is one of the many gifts of oil. It is only due to this dense energy product that the wars drag on for years and each time more deadly weapons are introduced with carnage growing exponentially.

In fact, the very concept of World War is actualized by discovery and subsequent usage of oil. Without this fossil fuel, the war zones could not have spread globally.

The origins of World War I were complex and included many factors, including the conflicts and antagonisms of the four decades leading up to the war.

After the onset of the Great Depression of 1873 in Britain, the sun began to set on the British Empire. By the end of the 19th Century, British industrial excellence was headed for a decline. The

*In two years Germany will be manufacturing oil and gas enough out of soft coal for a long war. The Standard Oil of New York is furnishing millions of dollars to help.*
*(Report from the Commercial Attaché, U.S. Embassy in Berlin, Germany, January 1933, to State Department in Washington, D.C,)*

decline paralleled an equally dramatic rise of a new industrial Great Power on the European stage, the Germany. Germany soon passed England in output of steel, in quality of machine tools, chemicals and electrical goods. Beginning the 1880's a group of leading German industrialists and bankers recognized the urgent need for some form of colonial sources of raw materials as well as industrial export outlet. German goods were of a superior quality and still lacked markets. Much later, this ire amply reflects in Hilter's calling Great Britain a 'shopkeeper's nation'. With Africa and Asia long since claimed by the other Great Powers, above all Great Britain, German policy set out to develop a special economic sphere in the imperial provinces of the debt-ridden Ottoman Empire. The policy was termed "penetration pacifique" an economic dependency which would be sealed with German military advisors and equipment. Initially, the policy was not greeted with joy in Paris, St. Petersburg or London, but it was tolerated. At the heart of this expansion policy laid the Berlin-to-Baghdad railway project, a project of enormous scale and complexity that would link the interior of Anatolia and Mesopotamia (today Iraq) to Germany. What Berlin and Deutsche Bank did not say was that they had secured subsurface mineral rights, including for oil along the path of the railway, and that their geologists had discovered petroleum in Mosul, Kirkuk and Basra.

The immediate origins of the war lay in the decisions taken by statesmen and generals during the July crisis of 1914, the spark for which was the assassination of Archduke Franz Ferdinand of Austria-Hungary by a Serbian irredentist. The crisis did not however exist in a void; it came at the end of a long series of diplomatic clashes between the Great Powers in the decade prior to 1914 which had left tensions high almost to a breaking point. In turn these diplomatic clashes can be traced to changes in the balance of power in Europe since 1870. But if we were to blame one single object for the scale of catastrophe, it was petroleum without which war's scope would have been limited in terms of time and space.

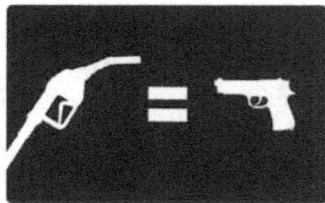

The belief that a war in Europe would be swift, decisive and "over by Christmas" is often considered a tragic underestimation; if it had been widely thought beforehand that the war would open such an abyss under European civilization, no one would have prosecuted it. Ivan Bloch, an early candidate for the Nobel Peace Prize, had predicted that an industrial warfare would lead to bloody stalemate, attrition, and even revolution on scales never witnessed before.

The one single factor that changed the course of war was British access to Iranian oil through a company known as Anglo-Iranian Oil.

In the first World War, in 1916, the combustion engine changed all the rules of the game. Fueled by petroleum, combustion engine increased mobility on the battlefield, spreading the conflict over a far greater area than any one could had ever imagined. War dynamics, dating back to thousands of years, overnight changed. The new fighting paraphernalia overwhelmed the best of infantry and cavalry. All because in the veins of these war machines flowed the potent liquid, petrol.

Over 13 million people perished and millions more were wounded during the four-year conflict.

Air warfare was also a new addition. Towards end of the war, Britain, Germany and France had ended up producing more than 1,50,000 planes. This war also saw the introduction of tank for the first time, and also cars, trucks and motorcycles.

It took huge quantities of oil to supply both sides' war effort. Oil production at Iranian wells was increased ten fold. Britain, captured Baghdad to further boost the supplies. Still France and Britain faced an acute shortage.

Germany's oil problems were worse. Allied naval blockade had cut off its supplies, leaving it with only one other option- the oil fields of Romania. Germans tried their best to capture Romanian fields but British army ahead of it, sabotaged the oil machinery. November 11, 1918 marked the day when a desperate Germany, faced with a serious oil shortage for the winter ahead, surrendered.

The decisive relationship of war and oil first emerged in the First World War. Britain, with its colonial control over Iranian oil, had a decisive advantage over the German-led Axis powers, allowing the Allies to "[float] to victory on a wave of oil," in the words of Britain's Foreign Secretary Lord Curzon.

## Post-War Explosions

In countless ways, World War I created the fundamental elements of 20th century history. Genocide emerged as an act of war. So did the use of poison gas on the battlefield. The international system was totally transformed. On the political right fascism came out of the war; on the left a communist movement emerged backed by the Soviet Union. Foundation was laid for America to become a world power. Of course until 1892, the United States was a second grade state at best and not even considered enough a contender at the table to warrant posting a full Ambassador level diplomatic mission. It was hardly played any serious role in European or Eurasian affairs. The Great Powers included Great Britain, France, the Austro-Hungarian Empire and Russia. After its defeat of France in 1871, Germany too joined the ranks of the Great Powers, albeit as a latecomer. After the world war 1, The British Empire reached its high point and started to unravel. Britain never recovered from the shock of war, and started her decline to the ranks of the second-class powers. At the peace conference of 1919, the German, Turkish, and Austro-Hungarian empires were broken up.

One thing was all too obvious after the World war 1- have oil or perish. There started frantic search for oil in different parts of the world. New fields were discovered in US, Mexico etc, but cynosure of all eyes was Middle East. Before the great war, US was a bystander as far as Middle East was concerned but now it too developed a keen interest in the region. Likewise there was much pushing and pulling, diplomatic maneuvers and once again the time arrived for another war.

## Second Great Oil War

Just as the war was ending, German Nationalists like Hitler gathered millions who rejected the peace and blamed Jews and Communists for their defeat. The road to the Second World War started there.

The Standard Oil group of companies, in which the Rockefeller family owned a one-quarter (and controlling) interest, was of critical assistance in helping Nazi Germany prepare for World War II. This assistance in military preparation came about because Germany's relatively insignificant supplies of crude petroleum were quite insufficient for modern mechanized warfare; in 1934 for instance about 85 percent of German finished petroleum products were imported. The solution adopted by Nazi Germany was to manufacture synthetic gasoline from its plentiful domestic coal supplies. It was the hydrogenation process of producing synthetic gasoline and iso-octane properties in gasoline that enabled Germany to go to war in 1940 — and this hydrogenation process was developed and financed by the Standard Oil laboratories in the United States in partnership with I.G. Farben.

Evidence presented to the Truman, Bone, and Kilgore Committees after World War II confirmed that Standard Oil had at the same time "seriously imperiled the war preparations of the United States."(Elimination of German Resources,p. 1085).

During World War II Standard Oil of New Jersey was accused of

*"The great two wars began from Europe simply on this basis. The German and Englishmen. The Englishmen, by their colonization, they made the whole world red in the map. Africa and Asia, India and America, Canada. And the Germans thought, "So this shopkeepers' nation..." Hitler used to say "shopkeepers' nation." "How they have occupied the whole world, and we are so intelligent? We are manufacturing so many things. We have no market to sell." That is the cause of the two great wars. This is a fact. Anyone, any politician, any gentleman knows what was the cause. Because England wants to lord it over, send Lord Clive to India to exploit. And the German wants that "We have got so many things manufactured. We cannot sell." That is the cause of war: lord it over. Everyone is trying to lord it over."*

*~Srila Prabhupada (Lecture, London, August 15, 1971)*

treason for this pre-war alliance with Farben, even while its continuing wartime activities within Himmler's Circle of Friends were unknown. The accusations of treason were vehemently denied by Standard Oil.

By the Second World War, the scramble for oil was a strategic priority on all sides. "The Japanese attacked Pearl Harbor to protect their flank as they grabbed for the petroleum resources of the East Indies," author Daniel Yergin wrote in his history of oil titled The Prize. "Among Hitler's most important strategic objectives in the invasion of the Soviet Union was the capture of the oil fields in the Caucasus. But America's predominance in oil proved decisive, and by the end of the war, German and Japanese fuel tanks were empty."

In this way, twenty years after the end of World War I, once again all the world powers knew that oil would make the difference between victory and defeat. The US once again used its plentiful oil fields to supply its own war needs and those of Great Britain. But Germany, just as in the First World War, was forced to undertake a risky strategy to capture foreign oil supplies.

Germany had two goals in mind - first capture Baku oil fields and then head for Middle East. A cut off fuel supply lines was synonymous with defeat.

Fuel shortages continued to haunt both sides as the war went on. General Rommel of the Afrika Korps, the best tank division Germany possessed, had to retreat in North Africa when the allies destroyed his refueling lines of supply. 'My men can eat their belts' he said, 'but my tanks gotta have gas.' was the comment made by General George Patton when faced with reversals.

Japan was very poor in most of natural resources, and it had to rely on import of these resources to function as a modern state. Among many natural resources, oil was one of the most crucial strategic materials that Japan desperately needed. Japan could not produce oil, within its borders, even for 10% of its domestic consumption. At

*"Violence is the inevitable result of attempting to achieve limitless growth in a finite world. If you want the limitless growth to continue, you must accept both the responsibility and the culpability of this never-ending violence."*
*- Stephen Hren*

24

the time, Japan had relied very much on the US, which supplied Japan about 80% of oil that was consumed in the island-nation. In other words, the power of life or death was in the hand of the American president. And President Roosevelt decided to choke Japan by keeping all oil, not even one drop, from going to Japan to make it comply with the demand of the US. Japan had also tied an economic treaty with Netherlands, which promised Japan the supply of oil (approximately 13% of the oil need) from the Dutch East India (Indonesia). However, the Dutch broke the treaty and followed the America's oil embargo in August of 1941 as well. That meant that there was no oil supply for Japan from the outside world, and the Japanese leaders had to find an alternative way to gain oil. Only option that the Japanese leaders could come up with was to take the Dutch East India and control oil fields. However, it was clear that if Japan just moved to south the war would become inevitable. The oil stock Japan had was only for a year and half, and time was running out. The Japanese leaders had to make up their minds as quickly as possible. If the war was unavoidable and they chose to fight, the longer they would wait the lesser the chance for victory would be because of the limited oil stock, which would be spent even during the peace time.

The final decision the leaders of Japan made was war, though most of them knew that their chance to beat the US was very slim, and on December 7th in 1941, the Japanese airplanes launched from the aircraft-carriers carried out a surprise attack on the US military bases in Pearl Harbor, Hawaii.

US was already on alert about Japan's imperial ambitions and had moved the American fleet from California to Pearl Harbor.

President Roosevelt had hoped that oil embargo would make them come to the bargaining table instead they came to pearl harbor, and Singapore etc.

Hoping to catch US by surprise they thought they may be able to buy time to expand deeper into western Asia to secure much needed oil, raw materials, and farmland. They believed that if they were going to spread westward they were going to have to disable the US fleet to give themselves a running start.

The December 7, 1941 attack destroyed a significant part of the fleet. But there was another blunder on their part. They failed to target the four and a half million barrels of oil stored at Pearl Harbor. This fuel saved American forces from getting completely immobilized in Pacific.

The turning point in the Pacific War was the battle of Midway in June 1942. From then on, the Allied forces slowly won back the territories occupied by Japan. In 1944, intensive air raids started over Japan. In spring 1945, US forces invaded Okinawa in one of the war's bloodiest battles. A key aspect of the US Navy's Pacific strategy was an intense campaign against Japanese commercial shipping. This blockade, primarily targeting oil, was spearheaded by US Navy submarines. A blockade proved the most effective means of attacking Japan's oil

With each defeat, Japan saw its access to oil dwindle. The US navy was sinking every Japanese oil tanker before it could return home from the Dutch East Indies and the Japanese Navy didn't have enough fuel to leave its home base. Oil, like that of many others, was instrumental in deciding Japan's fate too. Atom bombs made Japan surrender but it was oil paucity that defeated Japan by crippling its army and rendering its navy and air force completely useless.

Oil was the indispensable product, in all its forms, to the Allied campaigns around the world. Without it World War II could never have been won. For oil, once processed or refined in various ways, became the source or indispensable material for laying runways, making toluene (the chief component of TNT) for bombs, the manufacturing of synthetic rubber for tires and the distilling into gasoline for use in trucks, tanks, jeeps, and airplanes. And, that is not to mention the need for oil as a lubricant for guns and machinery.

To provide all the oil, or at least most of it, for the Allied war effort, the United States enlisted the aid of American oil companies, all of which responded without hesitation to the challenge. Meeting what everyone in government knew would amount to a demand for oil in unprecedented quantities required much organization. In May 1941, even then before the Japanese attack on Pearl Harbor, President Roosevelt established an official body known as Petroleum Administration for War (PAW).

American oil amounted in all to 6 billion barrels out of a total of 7 billion barrels consumed by the Allies for the period of World War II. Without this prodigious delivery, this global war might never have been won.

Field-Marshall Karl Rundstedt of Germany, when interviewed by newsmen, readily admitted how important oil had been in World War II. In fact, he attributed German defeat to three factors, :(1) the Allied bombing sorties (strategic and tactical); (2) the bombardments by Allied naval guns; and (3) Germany's own deficiency in oil, especially in the form of gasoline. The oil has not looked back ever since.

### Dawn of A Frenzied Oil Era

Allied forces were closing in on Nazi Germany and victory in Europe was just months away. For a week in early February 1945, U.S. President Franklin Delano Roosevelt met Winston Churchill and Stalin at Yalta to discuss the shape of post-war Europe. The summit ended on February 11, 1945 and Roosevelt departed for a rendezvous at the Great Bitter Lake, a waypoint along the Suez Canal in Egypt, with Saudi Arabia's King Ibn Saud, who sailed from Jeddah aboard an American warship to the meeting with Roosevelt. The two leaders' focus was shaping the future relationship between the United States and the Kingdom of Saudi Arabia. This meeting is considered very significant because just a few years back, a US President would not bother to give an appointment to spare few minutes to the king of a country of nomads. But now the issue was different - in 1939, Standard Oil of California along with Texaco had struck Saudi oil. The companies had agreed a 60 year concession with Ibn Saud covering 440,000 miles, one sixth of the continental US. Now, geologists at the newly formed Saudi-American Oil Company, Aramco, confirmed there was more oil underneath the Saudi sands than in the whole of the United States.

Officially President went to Great Bitter Lake because of oil. He was sailing back from his meeting at Yalta. It was a very dangerous time for him to be deviating from his path. War was not fully over yet and ships were very vulnerable. This explains the importance attached to this meeting. Roosevelt gave King Abdulaziz an airplane. The Americans were interested in the business arrangement and were not interested in culturally rearranging the country the way the British were known to do. They were not a colonial power. That meant an awful lot to the King, one of the only rulers in the area not colonized. He trusted the Americans in that they were unlike the British who were more meddlesome.

King clearly had a good grasp of geopolitics. He understood that the Americans were the up and coming international actors. He was more comfortable working with them in large part because of their lack of a colonial past. For those reasons he allowed the United States to build a base whereas the British had a much harder time getting access to the kingdom.

Thus was held the marriage of West and Middle East. US president Harry S. Truman wrote to the Ibn Saud in 1948, 'No threat to your Kingdom could occur that would not be a matter of immediate concern to the United States.' In next six decades, Americans and Saudis would have to confront imperialism, Nasserism, Communism, Baathism, Khomeinism, Islamic militantism, and terrorism but despite all this, the Saudi-American relationship would remain strong as long as oil flowed and Communism was fought.

World war II gave birth to two super powers - US and the Soviet Union and abruptly ended centuries old European domination. The Middle East oil was caught up in this struggle between two emerging super powers. Now if the Western world was going to survive economically, this oil, lying on the western side of the Iron Curtain would have to be protected. This was the foundation on which lay most of the post world war II diplomacies.

Demand for oil, all over the world and more so in US exploded in a post war era. This brought entire Middle East Asia into world focus. These obscure desert kingdoms rose to prominence and power and became craftsmen of world's destiny.

## Mideast (Land of Conflict) Occupies Center Stage

The Middle East is a term used to define a cluster of nations that include the Arab nations of South-West Asia, Israel and Egypt (geographically in Africa). Traditionally, the countries include Bahrain, Iraq, Iran, Israel, Jordan, Palestine, Kuwait, Lebanon, Oman, Qatar, Saudi Arabia, Syria, UAE and Yemen. The Middle Eastern nations can be divided into two groups. First group consists of rich nations with large petroleum supplies, relatively sparse population and with conservative regimes. Second group is made up of poor countries with little or no petroleum, heavily populated and having socialist governments. Islam is the common factor among all these countries with the singular exception of Israel, the only Jewish state in the world. Lebanon also is a multi-ethnic country, though the Hezbollah would like to transform it into an Islamic nation.

After World War II, Middle Eastern nations achieved independence. Colonial powers exited from the area. Formation of a Jewish state, Israel was an important event in post war Middle Eastern history. Jews had always claimed that Israel was their homeland dating back to thousands of years. But recent history reveals that Jews from all over the world began buying large farmlands from the Arabs in the last decade of 19th century. Israel then was ruled by the Turkish Ottoman empire but the area was won by Britain after World War I. Jews facing persecution in Germany, Russia and other parts of Europe began settling in Israel (then known as Palestine) in large numbers. Trouble between the Jews and the Muslims were a routine affair but with the passage of time as the Jews pressed their claim for a homeland, the rivalry turned into bitter enmity.

After the Holocaust, many of the survivors had no place to go. Many Jews (Zionists) believed that they should have a homeland of their own. They concentrated on the biblical area of Israel. After WWI, the area had become the British mandate of Palestine. When Jewish immigration accelerated, friction was created between Jews and Palestinian Arabs After 1945, Zionists and Palestinian Arabs wanted individual nations and both felt they had claim to Palestine. Britain withdrew in 1947 and the U.N. proposed that the country be partitioned 50/50. A war broke out when the Jews, certain of U.S. and Soviet support, declared their independence and the creation of a new state of Israel on May 14, 1948. When fighting ended in 1949, the Israelis had conquered more territory than had been envisioned in the U.N. plan, and the rest of the territory fell to Egypt and Jordan, rather than forming an independent Palestinian state. Palestinian Arab refugees fled to Lebanon, the West Bank, and the Gaza strip.

Post war Middle East era was marked by the superpowers trying to secure allies due to strategic importance of the area in the Cold War and vital petroleum fields. Middle Eastern Nations would devote large parts of their gross national product to large armies and arms purchases from both the West and the Soviet bloc.

Thanks to oil, the area remains one of the most volatile regions in the world. The region's history is tainted with skirmishes, conflicts and major wars.

A brief account of post world war major conflicts in middle-east is presented below.

### The 1948 Arab-Israeli War

Also known by Israelis as the War of Independence and by Palestinians as al Nakba ("the Catastrophe") was the first in a series of wars fought between the Israel and its Arab neighbors in the long-running Arab-Israeli conflict. The war marked the establishment of the State of Israel, and the exodus of hundreds of thousands of Palestinian Arabs from the territories that would become part of the new state.

### 1956 Suez War

The Suez Crisis was a military attack on Egypt by Britain, France,

> *"You can mostly forget ethnic or religious differences. The competition for a bigger share of the oil proceeds is behind much of the fighting."*
> *-Ed Harris*

and Israel beginning on 29 October 1956. The attack followed Egypt's decision of 26 July 1956 to nationalize the Suez Canal.

In 1952, army officers led a coup d'état against King Faruk and replaced him with President Gamal Abdel Nasser. Nasser became very popular in the Arab world and very unpopular in the West. Nasser nationalized the Suez Canal in 1956, leading to a war with Israel, France, and Great Britain.

Suez canal carried the majority of Middle East oil shipments to Europe even though stewardship of the canal was still controlled by Britain and France. In just a few months in 1955 Nasser successfully scared the hell out of US and Western Europe by turning to the Soviet bloc in search of weapons and raising the prospect that the canal might fall under Communist control.

Britain and France, fearing an economic catastrophe and also furious at the latest demonstration that their colonial power was fading out, took an aggressive step - they decided to invade the Suez to protect the canal. Israel, already looking for an excuse to topple Nasser, volunteered to join. The only ally they neglected to tell was a horrified US, not so much for the attempt to overthrow Nasser, but for the damage caused to Arab diplomacy.

Arab oil nations promptly banned all oil shipments to Britain and France and the US also declined to intervene. The Europeans had to retreat and with them retreated their influence in this oily region, region as volatile and slippery as oil.

### 1967 Six Day War

The Six-Day War, also known as the 1967 Arab-Israeli War, the Third Arab-Israeli War, Six Days' War, an-Naksah (The Setback), or the June War, was fought between Israel and Arab neighbors

Egypt, Jordan, and Syria. The nations of Iraq, Saudi Arabia, Kuwait, and Algeria also contributed troops and arms to the Arab forces.

1970 War of Attrition

The War of Attrition was a limited war fought between the Israeli military and forces of the Egyptian Republic, the USSR and the Palestine Liberation Organization from 1967 to 1970. It was initiated by the Egyptians as a way of recapturing the Sinai from the Israelis, who had been in control of the territory since the mid-1967 "Six-Day War". The hostilities ended with a ceasefire signed between the countries in 1970 with frontiers remaining in the same place as when the war began.

1973 Yom Kippur War

The Yom Kippur War, Ramadan War or October War, also known as the 1973 Arab-Israeli War and the Fourth Arab-Israeli War, was fought from October 6 to October 26, 1973, between Israel and a coalition of Arab states led by Egypt and Syria. The war began with a surprise joint attack by Egypt and Syria on the Jewish holiday of Yom Kippur. Egypt and Syria crossed the cease-fire lines in the Sinai and Golan Heights, respectively, which had been captured by Israel in 1967 during the Six-Day War.

1982 Lebanon War

The 1982 Lebanon War, called by Israel the Operation Peace of the Galilee and later colloquially also known in Israel as the First Lebanon War, began June 6, 1982, when the Israel Defense Forces invaded southern Lebanon. The Government of Israel ordered the invasion as a response to the assassination attempt against Israel's ambassador to the United Kingdom, Shlomo Argov by the Abu Nidal Organization.

**1987-1993 First Intifada**

The First Intifada (1987 - 1993) (also "war of the stones") was a mass uprising against Israeli military occupation, that began in Jabalia refugee camp and spread to Gaza, the West Bank and East Jerusalem.

Palestinian actions took a number of forms, including civil disobedience, general strikes, boycotts on Israeli products, graffiti, and barricades, but it was the stone-throwing demonstrations by youth against the heavily-armed Israeli Defense Forces that brought the intifada international renown.

Over the course of the first intifada, an estimated 1,100 Palestinians and 160 Israelis were killed in the fighting. Another 1,000 Palestinians were assassinated by their own people as alleged collaborators.

## 1982-2000 South Lebanon conflict

During the 1982–2000 South Lebanon conflict Hezbollah waged a guerrilla campaign against Israeli forces occupying Southern Lebanon. It ended with Israeli withdrawal in accordance with 1978's United Nations Security Council Resolution . Given that prior Arab Israeli wars were characterized by either Israeli victory or UN-enforced ceasefire, this is often regarded as a success of Hezbollah, which was able to extend its control of Southern Lebanon.

### Iran-U.S. Hostage Crisis (1979-1981)

On November 4, 1979, radical Iranian students seized the United States Embassy complex in the Iranian capital of Tehran. The immediate cause of this takeover was the anger many Iranians felt over the U.S. President Jimmy Carter allowing the deposed former ruler of Iran, Shah Reza Pahlavi, to enter the U.S. for medical treatment. In Iran, this was believed to be an opening move leading up an American-backed return to power by the Shah. The crisis which followed this seizure created a near state of war, ruined Jimmy Carter's presidency, and began an environment of hostility between America and Iran which continues to this day.

Ever since oil was discovered there in 1908, Iran had attracted great interest from the West. The British played a dominant role there until World War II, when the Soviet Union joined them in fighting to keep the Germans out. Until 1953, the United States mostly stayed on the sidelines, advocating for an independent Iran under the leadership of the young king, Reza Shah Pahlavi. But that year, fearing that charismatic prime minister Mohammed Mossadegh might be moving Iran closer to Moscow, the CIA directed an operation to oust him and consolidate power under the Shah.

With a steady flow of oil from the ground and military equipment from the U.S., the Shah led Iran into a period of unprecedented prosperity. But growing resentment against an uneven distribution of wealth and the westernizing influence of the United States led to a confrontation with Islamic clergy in 1963. The Shah effectively put down the uprising, sending its leader, an elderly cleric named Ruhollah Khomeini, into exile in Iraq. Though no one knew it at the time, Iran's Islamic revolution had begun.

Though fear of an American-backed return by the Shah was the publicly stated reason, the true cause of the seizure was the long-standing U.S. support for the Shah's government. Reza Pahlavi ruled Iran from 1941 to 1979, with a brief period of exile in 1953 when he fled to Italy due to a power struggle with Prime Minister Mohammed Mossadegh. Because Mossadegh's policies and announcements created concern over access to Iranian oil, oil prices, and possible Soviet influence in Iran, the United States and British intelligence services aided Iranian military officers in a coup to overthrow the Prime Minister. After his return to power, the Shah established a very close alliance with the United States. The U.S. supplied weapons, training, and technical knowledge that aided the Shah in modernizing his country. However, the Shah ruled as a dictator, using SAVAK, his secret police, to terrorize his political enemies. The Shah was opposed by both the Marxist Tudeh Party, and by fundamentalist Islamic leaders who believed his policies and his reliance on the Americans were corrupting Iranian society.

By 1978, unrest against the Shah had escalated into a violent uprising against his authority called the Iranian Revolution or the Islamic Revolution. On January 16, 1979, the Shah fled into exile for a second time, traveling to various countries before finally entering the U.S. for cancer treatments in October, 1979. After the Shah's departure, the Ayatollah Ruholla Khomeini returned from his own exile in France to take power over Iran. Khomeini was a leading member of the Shia Muslim clergy. The Shia are a subset of the Islamic faith, and form the majority of the Iranian population. Vital parts of this Islamic Revolution were propaganda and demonstrations against the United States and against President Jimmy Carter. After the Shah's

entry into the U.S., the Ayatollah Khomeini called for anti-American street demonstrations. On November 4, 1979, one such demonstration, organized by Iranian student unions loyal to Khomeini, took place outside the walled compound housing the U.S. Embassy.

Members of these Iranian student unions scaled the walls of the U.S. Embassy on November 4, 1979, taking 63 Americans hostage. Three more U.S. citizens were taken prisoner at the Iranian Foreign Ministry, for a total of 66 hostages. Within three weeks, the hostage-takers released several women and African-Americans, leaving 53. A sick hostage was later released, reducing the number to 52. Throughout their captivity, the hostages were paraded in front of television cameras, often blindfolded or hooded. Though the hostage-takers were not members of the Iranian government or military, their obvious, publicly-stated loyalty to Khomeini and the Islamic government created an international crisis.

Immediate official American reactions involved halting oil exports from Iran, expelling many Iranians living in the U.S., and freezing Iranian government assets and investments. Many Americans called for military action to free the hostages, but the situation became much more complicated when the Soviet Union invaded Iran's neighbor, Afghanistan, in order to crush an Islamic-based rebellion against that nation's Marxist government. President Carter now faced a crisis with oil-rich, but hostile Iran, a new Cold War crisis with the Soviets, and a growing sense in his own country that he was increasingly showing himself to be an ineffective leader.

Partly to counter the criticisms against him, as well as to free the hostages, President Carter ordered a military rescue mission code-named "Operation Eagle Claw." This mission was a total and complete failure resulting in the deaths of eight U.S. military personnel. On April 24, 1980, units of the rescue force landed in the Iranian desert to refuel their aircraft before heading to Tehran. A confusing series of events took place at this refueling point, including failed equipment, and desert sandstorms which reduced visibility. As a result of these problems, the rescue was called off. During the retreat, one of the helicopters collided with a transport airplane, causing an explosion which killed eight members of the rescue mission. Several of the burned American bodies were later part of grisly street demonstrations protesting the abortive U.S. "invasion" of Iran. A second rescue attempt was planned but never implemented, largely due to equipment failure.

On July 27, 1980, the former Shah died. Then, in September, 1980, President Saddam Hussein of Iraq invaded Iran. These two events led the Iranian government to enter into negotiations with the U.S., with Algeria acting as a mediator.

Domestically, the Hostage Crisis ruined President Carter's presidency. Unfortunately for him, the one-year anniversary of the embassy takeover fell on the same day as the United States Presidential election of 1980. Carter lost that election to former California Governor Ronald Reagan, who, though never publicly criticizing Carter over the hostage crisis, promised to rebuild American power and influence in the world.

The negotiations between Iran and the U.S. culminated in a deal that released the hostages and the eight billion dollars worth of frozen Iranian assets. Moments after Ronald Reagan took the oath of office on January 20, 1981, the hostages were allowed to fly out of Iran after 444 days of captivity.

The legacy of this prolonged crisis continue to affect Iranian-U.S. relations over 25 years later. Iran and the U.S. still do not have official diplomatic relations with each other, and both nations hurl hostile accusations at each other over issues such as the American invasion of Iraq and Iranian nuclear research. The June, 2005 election of Mahmoud Ahmadinejad as President of Iran opened up old wounds. Several of the former hostages contend that Ahmadinejad was one of the leaders of the student groups that seized the embassy in 1979. The Iranian government denies he had anything to do with the seizure, but it is another sign that tensions remain over the entire episode.

*"Oil is like a wild animal. Whoever captures it has it."*
*~J. Paul Getty*

## 1980-1988 Iran-Iraq war

The region has witnessed many wars, the biggest being the eight-year war between Iran and Iraq (September 1980 to August 1988). The battle of attrition was fought on the ground and in the air, and Iraq even used chemical weapons.

Iraq started the war by claiming the Shatt al-Arab (a 200 km river formed by the confluence of the Euphrates and the Tigris in the town of al-Qurnah in southern Iraq. Iraq was supported by the kingdoms of Saudi Arabia, UAE and Kuwait. Iran was backed by Syria and Libya.

There were no clear losers or gainers in the war, though both sides gutted many of each other's oil facilities.

It was of great cost in lives and economic damage - more than a million Iraqi and Iranian soldiers as well as civilians are believed to have died in the war with many more injured and wounded.

## The Soviet Invasion of Afghanistan (1979-1989)

Afghanistan had remained one of the poorest and least developed nations. Following a cycle of coups and countercoups, Babrak Karmal emerged and was backed by the Soviets. In December 1979, Soviet Union sent 80,000 troops to support the regime. Armed resistance by militant Muslims (Mujahedin) received support & training from U.S. Over a million refugees fled to Pakistan The mountainous terrain was ideal for guerrilla warfare and Soviet forces could not eradicate Afghan opposition The Soviets withdrew in 1988-89.

## The Gulf War (1990) - Iraq's Invasion of Kuwait

Iraq invaded the Kuwait (both were friends in the Iran-Iraq war) in August 1990 on a rather flimsy ground. It alleged that Kuwait was drilling oil in manner that sucked oil from under Iraqi territory!

*"What a country calls its vital economic interests are not the things which enable its citizens to live, but the things which enable it to make war. Petrol is much more likely than wheat to be a cause of international conflict."*
*~Simone Weil*

The war however was a short-lived after the US-led coalition backed Kuwait. The war was shown live on the CNN. The troops of many nations, from far and wide fought along side to defeat the Iraqi forces. Among them was the Islamic state of Pakistan. Most Middle East Arab nations fought against Iraq. These included Iraq's allies during the Iran-Iraq war, namely Saudi Arabia and UAE. Others like Syria, Turkey, Oman also were the part of the coalition that helped in driving the Iraqis out of Kuwait. Many non-Arab nations like France, Italy, Libya, Denmark, Canada, New Zealand also fought against the Iraq.

India didn't take part in the war; perhaps thinking that staying neutral was the best option, but it did extend to the USA refueling facilities for its aircrafts. Iraq was made to retreat within a 100 days.

## Organization Of Petroleum Exporting Countries (OPEC)

### An Icing On The Cake Of Conflict

The Organization of the Petroleum Exporting Countries (OPEC) was created at the Baghdad Conference in Iraq in September 1960. The founding members of the organization were Iran, Iraq, Kuwait, Saudi Arabia and Venezuela. These five states were later joined by eight other countries: Qatar (1961), Indonesia (1962), Libya (1962), United Arab Emirates (1967), Algeria (1969), Nigeria (1971), Ecuador (1973), and Gabon (1975). Ecuador and Gabon withdrew from the organization in 1992 and 1994, respectively.

The purpose of OPEC, as with any cartel, is to limit/regulate supplies in the hope of keeping prices high. The oil industry has been plagued by production booms and falling prices ever since Colonel Drakes' discovery of oil in Pennsylvania in 1859. Just as the major oil companies colluded from the 1920's to the1960's to prevent prices (and profits) from falling, members of OPEC meet on a regular basis to set production levels in the hope of maintaining prices. The essential nature of oil (no substitutes) coupled with its limited number of suppliers make it the ideal product for cartelization.

The rise of OPEC is tied to a shifting balance of power from the multinational oil companies to the oil producing countries. Lacking exploration skills, production technology, refining capacity, and distribution networks, oil producing countries were unable to challenge the dominance of the oil companies prior to World War II. Although Mexico wrested control of its oil industry from foreigners in 1938, it quickly receded from the lucrative international market due to insufficient capital for investment.

However, about the time of World War II the oil exporting countries began seeking better terms in their oil contracts. In 1943 Venezuela signed the first "fifty-fifty principle" agreement which provided oil producers with a lump sum royalty plus a fifty-fifty split of profits (i.e., selling price minus production cost).

In the late 1940's Venezuela revised their tax system to capture a greater share of the oil profits. The oil companies responded to this move by shifting oil purchases to countries with cheaper contracts. In response, Venezuela contacted Arab producers and encouraged them to demand similar "fifty-fifty" deals and reform their tax systems. Saudi Arabia, seeing the value of the fifty-fifty contract and understanding the power of acting collectively, quickly demanded and received a similar contract from Aramco.

In 1947, the Iranian Parliament passed a law demanding the termination of previous agreements with Anglo-Iran (referred to as Anglo-Persian prior to 1935 and British Petroleum after 1954). When negotiations failed to lead to a compromise, Iranian Prime Minister Mossadegh nationalized oil operations in May 1951. The collapse of the oil industry pushed the economy into chaos.

Domestic opponents, aided by the American Central Intelligence Agency, were able to topple Mossadegh in 1953. A new British-Iranian agreement was signed the following year. The newly restored Shah of Iran became a pillar of American middle east policy until the Iranian Revolution in 1979.

While world oil demand grew during the 1950s, they were outpaced by the growth in production. The problem was exacerbated by the fact that the "fifty-fifty" deals were based on "posted" prices rather than "market" prices. Given that posted prices were fixed, oil producing countries had an incentive to grant additional concessions to expand oil revenue. Market prices became divorced from their calculations. The increases in supply drove market prices even further down and eroded the profits of the multinational oil companies.

The downward push on prices led to a policy debate in Washington. Although the United States had been a net exporter of oil until 1948, the expansion of cheaply produced oil from the middle east led to rising imports. As prices fell, domestic producers simply could not compete. Moreover, the Eisenhower Administration concluded (as the Japanese had prior to World War II), dependence on foreign oil placed the country's national security in jeopardy. The U.S. responded with an import quota. The quota kept domestic prices artificially high and represented a net transfer of wealth from American oil consumers to American oil producers. By 1970, the world price of oil was $1.30 and the domestic price (U.S.) of oil was $3.18 (Danielsen 1982, 150).

During its first decade, OPEC was able to halt the free fall in prices. However, it was not able to raise prices as most members had hoped. In general, commodity cartels (such as the tin cartel or the coffee cartel) collapse because there are many substitutes for the product or there are many potential producers of the product. A cartel-inspired rise in coffee prices triggers some consumers to switch to hot tea (i.e. demand falls) and encourages new producers to enter the market (i.e., supply rises). Both the fall in demand and the rise in supply put downward pressure on prices and undermine the cartel's effectiveness. Cartels also suffer from a "collective action problem." That is, every member has an incentive to cheat on the cartel by increasing its production. For example, an individual country such as Iran can increase its oil revenues by expanding production as long as all other members stick to their quotas. However, all members have a similar incentive to increase production — i.e., they all want to free ride on the collective good. The incentive to cheat implies that cartels are traditionally short-lived enterprises.

Although the essential nature of oil and the limited number of suppliers worked in OPEC's favor, the power of the organization remained limited during the first decade for four reasons. First, OPEC's share of world production was only 28% in 1960. By 1970, this figure would rise to 41%. Second, the fact that the oil reserves in the ground belonged to the multinational corporations (except in Iran) limited the power of the oil producing countries. Third, the oil exporting countries were desperate for revenue to fuel economic development. Fourth, important political divisions existed in the Arab world. The revolutionary government of Nasser repeatedly clashed with the Saudi monarchy. Iraq threatened to invade its neighbor Kuwait (it was deterred by the deployment of British forces). Iran and Saudi Arabia vied for leadership of the Middle East.

## The First Oil Shock

OPEC's fortunes began to shift in the early 1970's as rising demand for oil began to outstrip production. Moreover, the oil- producing states began demanding further concessions. Muammad al-Qaddafi, after seizing power in military coup in Libya, demanded and received a 20 percent increase in royalties, a "55-45" profit sharing agreement, and tax concessions (Yergin 1991, 580). This move triggered a series of new demands that ratcheted up oil prices and oil-exporting countries' profits.

As the world oil market tightened, the Arab world became more vocal in calling for use of the oil weapon to achieve their economic and political objectives. This was most acutely realized in the oil embargo during the 1973 October War between Egypt and Israel. Saudi Arabia refused to increase production in order to halt rising prices unless the U.S. backed the Arab position. Arab oil ministers than agreed to an embargo to further their political objectives. Production would be cut by 5 percent per month until the West backed down. States adopting a "friendly" position (from the Arab

*Operation Iraqi Liberation, O.I.L.—OOPS! That pesky truth tries to wriggle out every time we let our propaganda slip. O.E.F., now that sounds much better, Operation Enduring Freedom.*
*~ Rand Clifford*

perspective) would be unaffected. When Nixon publicly proposed a $2.2 billion military aid package for Israel, Arab states began an oil embargo against the United States (later expanded to the Netherlands, Portugal, South Africa, and Rhodesia).

The new official price was agreed among OPEC member countries: $11.65. Oil prices jumped from about $3.00 a barrel before the war to $11.65. The embargo, which did not end until the Syrian-Israeli disengagement was secured, drove the world economy into deep recession. Gross national product in the U.S. declines by 6 percent in the following two years. The Japanese economy shrinks for the first time since the Second World War

## The Second Oil Shock

The Second Oil Shock began when the Iranian Revolution and ensuing halt of Iranian petroleum exports had caused panic and speculations in the world oil market. When the Carter administration placed an embargo on the importing of Iranian oil into the United States and froze Iranian assets in response to the hostage taking, Iran counter-attacked by prohibiting the exporting of Iranian oil to any American firm.

Moreover, the outbreak of the war between Iran and Iraq in 1980 shook the oil market as well. In its initial stage, the Iran-Iraq war abruptly removed almost 4 million daily barrels of oil from the world market—15 percent of total OPEC output and 8 percent of free world demand. In 1980 OPEC representatives (with the exception of Saudi Arabia) agreed to set prices at thirty-six dollar a barrel.

*"9/11 gave the Bush people carte blanche to carry out their extreme agenda – and they didn't hesitate for a moment to use it... They've used the war on terror to justify everything from tax cuts to Alaska oil drilling."*
*~Ron Reagan*

However, the impact of the Second Oil Shock turned out to be short-lived. The influence of OPEC appeared to be diminishing as the production by Mexico, Britain, Norway, and other non-OPEC countries and Alaska was continuing to increase. Anxious to increase market share, they were making significant cuts in their official prices. As a result, OPEC's share of world output quickly fell by 27 percent. Oil revenues for OPEC members plunged after 1981. Saudi Arabia, the largest producer in OPEC, saw its oil revenues plunge from $113.2 billion in 1981 to just $20.0 billion in 1986.

Although the Second Oil Shock sent the developed world into recession, the most serious long run impact of the second shock was in the developing world. During the 1970s, the oil producing states placed a significant portion of their revenue into commercial banks because they simply could not spend the money as fast as it came in. The commercial banks loaned this money to developing countries which hoped to repay the loans with revenue from their rapidly growing economies. However, the developed world responded to the Second Oil Shock by rapidly raising interest rates which deepened the on-going recessions. The developing countries saw exports fall, oil import prices rise, and interest payments skyrocket. The result was the debt crisis which first appeared in Mexico in 1982 and quickly spread throughout the developing world. In the "lost decade" of the 1980's, years of hard fought economic gains were wiped out. From 1980-88, the real income of American workers fell by 40 percent (Lairson and Skidmore 1993, 277).

## The Third Major Price Spike

The third major price spike occurred in 1990-91 when Iraq invaded its fellow OPEC member Kuwait. Iraq had long claimed the territory of Kuwait; in 1961 it appeared Iraq was going to swallow its tiny neighbor until the dispatch of troops by the British. In 1991

*"As we head to war with Iraq, President Bush wants to make one thing clear: This war is not about oil, it's about gasoline."*
~Jay Leno

the on-going territorial conflict was exacerbated by two oil issues: (1) the continued pumping of oil by Kuwait from a field located under both countries; and (2) low oil revenues for Iraq which made paying off its war debts (to Kuwait and others) difficult. A successful invasion would expand reserves, augment Iraqi power in OPEC, raise oil prices and revenue, and annul war debts to Kuwait.

Iraq gambled that the U.S. response would be political and economic. However, the Iraqi invasion triggered a military response which was supported by an unlikely coalition of western, developing, communist, and Arab states. The sudden removal of two major producers, Kuwait and Iraq, could have sent oil prices through the ceiling. However, Saudi Arabia expanded production by literally millions of barrels per day to keep prices from rising a great deal. Since the war, Iraq's refusal to comply with United Nations resolutions had resulted in the continuation of an oil embargo.

## The Future of OPEC

Oil prices are rapidly rising and are at an all time high. It is clear that demand for oil will continue to rise. Moreover, the lack of major oil finds in the last twenty years implies that supply will grow only slowly, if at all. Unless a cheap alternative source of energy is discovered in the meantime, this combination of effects will create an environment conducive to cartelization.

## Current Day Conflicts

War is now incessant and ubiquitous, but its nature has changed. Conflicts are, not so much between nations, but rather between ethnic, linguistic, and religious groups.

Today's conflicts are not arising from head-on clashes between the great powers, but rather through the escalation of local conflicts sustained by great power involvement. Today's great powers -- led by the United States and China -- are developing or cementing close ties with favored suppliers in the Middle East, Central Asia, and Africa. In many cases, this entails the delivery of large quantities of advanced weaponry, advisors, and military technology -- as the United States has long been doing with Saudi Arabia, Kuwait, and the United Arab Emirates, and China is now doing with Iran and Sudan.

Nor should the possibility of a direct clash over oil and gas between great powers be ruled out. In the East China Sea, for example, China and Japan have both laid claim to an undersea natural gas field that lies in an offshore area also claimed by both of them. In recent months, Chinese and Japanese combat ships and planes deployed in the area have made threatening moves toward one another; so far no shots have been fired, but neither Beijing nor Tokyo have displayed any willingness to compromise on the matter and the risk of escalation is growing with each new encounter.

The likelihood of internal conflict in oil-producing countries is also destined to grow in tandem with the steady rise of energy prices. The higher the price of petroleum, the greater the potential to reap mammoth profits from control of a nation's oil exports -- and so the greater the incentive to seize power in such states or, for those already in power, to prevent the loss of control to a rival clique by any means necessary. Hence the rise of authoritarian petro-regimes in many of the oil-producing countries and the persistence of ethnic conflict between various groups seeking control over state-oil revenues -- a phenomenon notable today in Iraq (where Shiites, Sunnis, and Kurds are battling over the allocation of future oil revenues) and in Nigeria (where competing tribes in the oil-rich Delta region are fighting over measly "development grants" handed out by the major foreign oil firms).

When we reach the point where the world's oil-hungry economies are competing for insufficient supplies of energy, oil will become an even stronger magnet for conflict than it already is.

Oil considerations dominate and guide US foreign policy decisions. Therefore, as a result of this realisation, if one were to entertain some lurking doubts about stated US public policy – democracy, freedom, etc – that could be understandable, after reconsidering US military actions in Afghanistan and Iraq. In contrast and in contradiction to an apparent benign and enlightened foreign policy rhetoric, its militaristic adventures are simply neo-colonial wars.

## Terror Chant

This numerical phrase, 9/11 and the catch-all phrase "war on terror" have repeatedly been recited and relied on by the Bush administration to justify military action in Afghanistan and Iraq as well as the imposition of other draconian measures on some of its own citizens, e.g. the Patriot Act, and other homeland security measures. However, it has now been clearly demonstrated in all manner of ways, and even by Bush's belated grudging acknowledgment – that before the US invasion of Iraq there was no training of, or support for terrorists in Iraq; that Iraq was not intent on attacking the US. WMDs (Weapons of Mass Destruction) non-existence speaks volumes about a lying and deceptive US administration. The term "terrorism" has now become a fashionable tool or a tactic that some unscrupulous countries are now using, taking their cue from the US, to crush or suppress any legitimate dissent or opposition within or outside their borders.

Consider these statements: -

"Hussein has not developed any significant capability with respect to weapons of mass destruction. He is unable to project conventional power against his neighbors."

*–Colin Powell on February 24, 2001*

*"we are entering a historical period of potentially great instability, turbulence and hardship. Obviously, geopolitical maneuvering around the world's richest energy regions has already led to war and promises more international military conflict. Since the Middle East contains two-thirds of the world's remaining oil supplies, the U.S. has attempted desperately to stabilize the region by, in effect, opening a big police station in Iraq. The intent was not just to secure Iraq's oil but to modify and influence the behavior of neighboring states around the Persian Gulf, especially Iran and Saudi Arabia. The results have been far from entirely positive, and our future prospects in that part of the world are not something we can feel altogether confident about."*

*- James Howard Kunster, The Long Emergency*

"Simply stated, there is no doubt that Saddam Hussein now has weapons of mass destruction,"

*–Dick Cheney on August 26, 2002.*

"Our conservative estimate is that Iraq today has a stockpile of between 100 and 500 tons of chemical weapons agent. That is enough agent to fill 16,000 battlefield rockets. Even the low end of 100 tons of agent would enable Saddam Hussein to cause mass casualties across more than 100 square miles of territory, an area nearly five times the size of Manhattan."

*–Colin Powell at the UN on February 5, 2003*

"Intelligence leaves no doubt that Iraq continues to possess and conceal lethal weapons."

*–George W. Bush on March 18, 2003*

"We are asked to accept Saddam decided to destroy those weapons. I say that such a claim is palpably absurd."

*–Tony Blair, Prime Minister 18 March 2003*

"Saddam's removal is necessary to eradicate the threat from his weapons of mass destruction."

*–Jack Straw, British Foreign Secretary 2 April 2003*

"Before people crow about the absence of weapons of mass destruction, I suggest they wait a bit."

*–Tony Blair 28 April, 2003*

These facts can not leave any sensible and rational person in doubt about the levels of dishonesty, collusion, fabrication and calculated

*"No one really wants to send their kids off to die for oil."*
*~Daryl Hannah*

deception that led the invasion of Iraq in 2003.

Decision makers in the US bombed, invaded, occupied and caused at lowest estimate over 30,000 innocent civilian deaths in Iraq's oil war. These are the human consequences of the way in which powerful interests pursue the ends of dominance over global oil supplies.

### From Cold War Era To Energy War Era

Speaking at a conference named "Summit on Energy Security" in US, the chairman of the US Senate Foreign Relations Committee, Richard Lugar, characterized Venezuela, Iran and Russia as "adversarial regimes" that were using energy supplies as "leverage" in foreign policy.

Lugar said: "We are used to thinking in terms of conventional warfare between nations, but energy is becoming a weapon of choice for those who possess it."

Senior Russian figures were quick to dismiss Lugar's admonition as "groundless Russophobia", but the US administration is already opening new battle fronts against Russia in the energy war.

With rivalry building up over the Caspian pipeline, Russia has been under pressure to find an alternative evacuation route for Kazakh oil. For Greece and Bulgaria too, picking the Russian proposal meant ignoring US entreaties for an alternative US-backed non-Russian

pipeline system that was already on the drawing board - an Albania-Macedonia-Bulgaria route for southwestern Europe.

Last year, while visiting Ankara and Athens, US Secretary of State Condoleezza Rice publicly warned Turkey and Greece about any collaboration with Russia that would facilitate Russia's tight grip on European energy supply. "It is quite clear that one of the [US] concerns is that there could be a monopoly of supply from one source only, from Russia," Rice said.

### Afganistan - Crucial Geostrategic Position

Thus 9/11 could be considered as horrific blowback from the CIA having funded, armed and encouraged Muslim fundamentalist and Taliban regime. However, with about 15 Saudis directly involved in the 9/11 attack the nexus of the attack raises even more questions beyond the assumption that one man stationed in remote Afghanistan, Osama bin Laden, almost unilaterally orchestrated an attack which US intelligence remained ineffective to stop. So Afghanistan and Iraq are essentially oil related military operations pursued by the US in an on-going oil-war,

There is a massive triangle within which the world's largest supplies of oil and natural gas are to be found. Within the area of this triangle are to be found regions and countries such as:

• The Caspian Sea (with surrounding countries Kazakhstan, Turkmenistan, Iran and Azerbaijan)

• Central Asia (including Kazakhstan, Uzbekistan, Tajikistan, Kyrgyzstan, Turkmenistan, Afghanistan, Pakistan and into China and India)

• The Persian and Arabian Gulf states (Oman, United Arab Emirates, Qatar, Saudi Arabia, Iraq and Iran)

These areas - this triangle of oil and natural gas - hold the world's greatest reserves of oil and natural gas, which are mirrored, in the global politics of oil.

The US bombed, and has occupied Afghanistan pursuant to a declared policy of pursuing Osama bin Laden. There is a complementary logic of US military occupation of Afghanistan. If the US is to become less dependent on Arab oil, its focus will be the oil and natural gas resources of Central Asia and the Caspian Sea regions. However, access to these alternative supplies of oil requires pipeline routes. Afghanistan's geographical position serves well, the oil and natural gas pipeline transit requirements for a route from Central Asia to the Arabian Sea. For the US to establish and keep the pipeline functional, Afghanistan will have to be politically tamed. This political taming of Afghanistan translates in military terms to having an occupational force in Afghanistan, effectively for the policing of the pipeline against sabotage and controlling the regime in Kabul to be accommodative to US oil interests. Viewed in this light, one can more realistically understand sustained US military action in Afghanistan. If Turkmenistan and Uzbekistan are to transit natural gas and oil independently of Russia, then stability in Afghanistan is vital from an oil geostrategic perspective. Although Afghanistan has very little oil, its strategic importance resides in being a country central for transit to port. US acceptance of Pakistan's dictatorship under Musharraf is better understood when one views Pakistan's location next to the Arabian Sea and Afghanistan's proximity next to Pakistan – how is the oil to reach port if not through Pakistan? In 2005 Russia briefly disrupted gas supplies via a pipeline transiting gas through Ukraine. This action demonstrates the kind of strategic calculations one would then encounter if significant supplies destined for the US had a reliance on lines that traversed geographical territory that was under Russian or other potentially threatening control.

*"So in the Vedic civilization there was no such problem as petrol problem and food problem or... no. The problem was whether the civilization was going nicely, whether the human civilization is making progress toward the ultimate goal of life, not to bother with the temporary problems."*
*~Srila Prabhupada (Lecture on Srimad-Bhagavatam, Hawaii, January 17, 1974)*

## Lessons Not Learned

Now the attention is focussed on Iran and there is a sudden and unexplained fear that it might develop some objectionable weapons that might pose a threat to someone in the future. North Korea, which already has nuclear weapons and long range missiles - and isn't exactly a friendly place - is not deemed a threat. The cynic can be forgiven for thinking there is some other motive for these military moves: could it be oil?

The Uranium Information Centre in Melbourne, Australia, confirms that there are 31 countries with nuclear power plants. There are a further 7 seeking to acquire nuclear capabilities. The US bellicose position on Iran, in attempting to justify another oil war against Iran would have to be seen in this light. A 32nd country having nuclear capabilities will not be a threat to the US.

The crux of matter here again, is oil.

## American Army : Global Oil-Protection Service

It has been argued that oil-protection role is a peculiar feature of the war in Iraq, where petroleum installations are strewn about and the national economy is largely dependent on oil revenues. But Iraq is hardly the only country where American troops are risking their lives on a daily basis to protect the flow of petroleum. In Colombia, Saudi Arabia, and the Republic of Georgia, U.S. personnel are also spending their days and nights protecting pipelines and refineries, or supervising the local forces assigned to this mission. American sailors are now on oil-protection patrol in the Persian Gulf, the Arabian Sea, the South China Sea, and along other sea routes that deliver oil to the United States and its allies. In fact, the American military is increasingly being converted into a global oil-protection service.

*"The oil axis is present in most of the U.S. administration, beginning with its president, vice-president, and top advisers, including Rice, who is oil-colored."*

## China, India Join The Race

In less than a year, India and China have managed to confound analysts around the world by turning their much-vaunted rivalry for the acquisition of oil and gas assets in third countries into a nascent partnership that could alter the basic dynamics of the global energy market. The position of Asia couldn't be more abject. A continent which hosts the world's largest producers and fastest growing consumers of energy is forced to play second fiddle, relying on institutions, trading frameworks and armed forces from outside the region in order to trade with itself. Such a situation makes for unstable politics and bad economics, not to speak of atrocious geography. Central Asia is close to China and Iran but the U.S. has spent the better part of a decade trying to make sure pipelines carrying oil and gas from there only go westward. Gas pipelines connecting Iran to India make financial sense but the threat of U.S. sanctions means this project might not get off the ground. If the 21st century is to be an 'Asian century', Asia's passivity in the energy sector is slated to end in near future and will thus further exacerbate the race.

While efforts are under way to seal nuclear deals with the US and France to generate electricity, India's efforts to tie up gas resources as another alternative to fossil fuels have gathered momentum.

Following the decision by Myanmar to supply gas to China, India is now making swift maneuvers to ensure that the US$1 billion Myanmar-Bangladesh-India (MBI) gas pipeline materializes. And significantly, India has virtually decided to join the US-backed Turkmenistan-Afghanistan-Pakistan (TAP) pipeline, in part because of the geopolitical difficulties involved in the $7 billion Iran-Pakistan-India (IPI) pipeline that Washington opposes.

## Oil & Terrorism - A Strategic Partnership

Terrorist have taken out 25 per cent of Nigeria's sweet crude since late February and the daily joust between Washington and Tehran is providing splendid returns to those who had invested in oil stocks.

Just as things seemed to be calming down in the delta region of Nigeria after a spate of kidnappings and insurgent attacks, the militant group calling itself the Movement for the Emancipation of the Niger

Delta — MEND — announced recently to all who would listen that it was planning new violence against oil facilities in the region.

If a rag-tag bunch of militants in Nigeria are accurately reading the tea leaves in an oil-brimmed cup, imagine what Al Qaeda and the pan-global martyrs of terror have in store?

There is a greater lead-filled premium now for oil-related and economic targets.

Terrorists are ready to throw their explosive spanners into the machineries of global trade. They know the drill. There is tremendous potential here to cause pandemonium and cross-border tensions. Targeting energy infrastructures will be far more efficacious and less scornful than the indiscriminate slaughter of civilians.

Economic targets offer the best returns with less risks. Stock markets will be depressed, inflation will set in, bankruptcies will proliferate, and people will starve.

There are Maoist guerillas in Nepal, ethnic insurgents in Burma, Al Qaeda-inspired militants in Islamic nations and assorted purveyors of violence in every nook and corner where there are power plants, ports, retail stores, trading depots, factories, banks and transborder pipelines.

If Iran burns, or if neighboring Iraq descends further into anarchy, expect scattered strikes against oil installations, ports and power plants the world over. There will be more trouble in Nigeria and Ecuador. Hotspots will get hotter with conflicts spreading far and wide.

If Iran gets hit, the anti-American Venezuelan President Chavez may deliberately divert oil supplies to nations like China, depriving the US of 15 per cent of its oil imports.

### Oil Discovery - A Curse For Africa

The United States and other developed countries are increasingly turning to West Africa in their scramble for oil, but for Africa the oil boom is like a disease that creates poverty, conflict and corruption.

That's the diagnosis of Nicholas Shaxson, an Africa expert whose book *Poisoned Wells - The Dirty Politics of African Oil* tours some of Africa's poorest and most violent hot-spots.

From simmering conflicts in the Niger Delta to civil war in Angola to rampant corruption in Gabon and Equatorial Guinea, Shaxson contends that these countries are worse off than they were before they struck it rich.

Some countries in Africa draw in m ore money from oil than they do from foreign aid but tend to get poorer and more violent over time, while their rulers jet off on shopping excursions to Paris.

54

## Section-III

# Peak Oil

### Dwindling Supplies

*"My father rode a camel; I drive a car; my son flies a jet; his son will ride a camel."*   *-Saudi saying*

# Thinking The Unthinkable

Carl Jung, one of the fathers of psychology, famously remarked that "people cannot stand too much reality." And one such reality is that we have to bid good bye to oil one day. Its not a question of what or if, but just a question of when. And the thing is we don't have to run out of oil to start having severe problems with industrial civilization and its dependent systems. We only have to slip over the all-time production peak and begin a slide down the arc of steady depletion. In other words, we won't have to run completely out of oil to be rudely awakened. The panic starts once the world needs more oil than it gets. The key event in the Petroleum Era is not when the oil runs out, but when oil production peaks. Just like they thought that the Titanic was unsinkable. The upcoming end of cheap oil seems to have surprised markets. The exponential increase in demand for fossil fuels seems to have come as an unpleasant surprise. The alternative sources of power: solar, wind, nuclear, tidal, etc. are not as energy dense, portable, or as readily usable as fossil fuels. History tells us that complete development of new energy sources (coal and oil in the past) takes a long time, at least about half a century. The peak in fossil energy extraction will expose the fallacy of limitless growth.

*Six Basic Indisputable Facts About Petrol*
*-There is a finite and limited reserve of oil in this world.*
*-End of oil is not a question of 'what' or 'if' but just a question of 'when'.*
*-There is great concern in political circles about continuity of oil supply.*
*-No alternative energy source comes even close to the convenience and economics of petrol and no major breakthroughs are in sight.*
*-Our lives are increasingly dependent of abundant and cheap supply of oil.*
*-Time to be conscious of these facts, leave aside acting on this, is now!*

## The Wolf Finally Arrives

### The First Ever Acceptance Of The Upcoming Crisis By The Oil Industry

Here we reproduce a press report dated 22nd July, 2007.

Oil And Gas May Run Short By 2015
By Geoffrey Lea

The Independent
22July, 2007

Humanity is approaching an unprecedented crisis when not enough oil and gas will be produced to keep industrial civilisation running, the world's top oilmen warned last week.

The warning – which is being hailed as a "tipping point" on both sides of the Atlantic – marks the first time that the industry has accepted that it may soon no longer be able to meet demand for its products. In Facing the Hard Truths about Energy, it gives authoritative support to concern about impending shortages, following a similar alert by the International Energy Agency less than two weeks ago.

The 420-page report, the most comprehensive study ever carried out into the industry, has been produced by the National Petroleum Council, a body of 175 authorities that reports to the US government. It includes the heads of the world's big oil companies including ExxonMobil, Chevron, ConocoPhillips, Occidental Petroleum, Shell and BP.

It is also remarkable for the conversion of its chairman, Lee Raymond, the recently retired chief executive of ExxonMobil, who led opposition against action to tackle global warming, and became environmentalists' most prominent bogeyman. The report argues for "an effective global framework" to manage emissions of carbon dioxide – "incorporating all major emitters" – and urges the US to cut the pollution that causes climate change.

The report concludes that "the global supply of oil and natural gas from the conventional sources ... is unlikely to meet ... growth in demand over the next 25 years". It says that "many observers think that 80 per cent of existing oil production will need to be replaced by 2030" to keep up present supplies "in addition to volumes required to meet existing demand." But, it adds, there are "accumulating risks to replacing current production and increasing supplies".

Though vast amounts of oil and gas remain underground, "complex challenges" and "global uncertainties" are likely to put an end to "the sufficient, reliable and economic energy supplies upon which people depend". And the crunch could come sooner, with oil production becoming "a significant challenge as early as 2015". This chimes with the International Energy Agency's prediction that oil supplies could become "extremely tight" in five years.

The predictions should send a shiver down humanity's collective spine as a shortage of oil and gas has been predicted to cause industrial collapse, market crashes, resource wars and a rise in poverty. Some forecast that fascist regimes will rise out of the chaos.

Chris Skrebowski, editor of the Energy Institute's Petroleum Review, said the report's publication showed the industry "'fessing up that it really has a problem on its hands". Until now, he said, "companies, full of share options, have been terrified of frightening the markets" by revealing the truth.

The report says the fuel efficiency of cars should be increased "at the maximum rate possible" and there should be a crackdown on 4x4s. It calls for "aggressive energy efficiency standards for buildings, and measures to "set an effective cost for emitting carbon dioxide" to combat global warming.

© *2007 Independent News and Media Limited*

*"If liberty means anything at all, it means the right to tell people what they do not want to hear."*
*~George Orwell*

# Facing A Wall of Uncertainty

Shihab-Eldin, secretary-general of OPEC said recently, "When we look at the future, we find ourselves facing a wall of uncertainty."

Oil is a finite resource, and there will come a day, inevitably, when we reach the highest amount of oil that can ever be pumped. Beyond that day - which we can think of as the topping point, or "peak oil" as it is often called - will lie a progressive overall decline in production. Putting the same question a different way, then, at the current prodigious global demand levels, where does oil's topping point lie?

This half-century of deepening oil dependency would be difficult to understand even if oil were known to be in endless supply. But what makes the depth of the current global addiction especially bewildering is that, for the entire time we have been sliding into the trap, we have known that oil is in fact in limited supply. At current rates of use, the global tank is going to run too low to fuel the growing demand sooner rather than later this century. This is not a controversial statement. It is just a question of when.

The scale of the addiction - and of the resource - is smaller. But the patterns are the same: growing demand for a finite resource, most of which has to be imported from the Middle East and the former Soviet Union. Even a temporary blip in supply is enough to create something close to panic among governments. It is oil that keeps our civilization functioning.

The world is running amok by gulping petroleum in an ever increasing way. America is not alone in her addiction and her dilemmas. Whole world is trying to follow in the footsteps of America. The motorways of Europe now extend from Clydeside to Calabria, Lisbon to Lithuania. Growing economies like India, China have joined the race.

# PEAK OIL

## The beginning of the end of oil

Peak oil means the end of cheap oil, and an end to economies organized around the increasing availability of cheap oil. As first expressed in Hubbert peak theory, peak oil is the point or timeframe at which the maximum global petroleum production rate is reached. After this timeframe, the rate of production will by definition enter terminal decline. According to the Hubbert model, production will follow a roughly symmetrical bell-shaped curve.

The peak is the top of the curve, the halfway point of the world's all-time total endowment, meaning half the world's oil will be left. That seems like a lot of oil, and it is, but there's a big catch: It's the half that is much more difficult to extract, far more costly to get, of much poorer quality and located mostly in places which are politically instable. A substantial amount of it will never be extracted.

In the 1950s, a prominent geologist M.King Hubbert found that as the years went by, U.S. domestic oil production was decreasing, mainly because new discoveries became fewer and smaller. The changes in production could be plotted on a graph, forming the left side of that familiar shape known as a bell curve. Looking at the graph, Hubbert could see that the peak of American oil production would be about 1970; after that, there would be a permanent decline. When he announced this, most people laughed at him. But he was right: after 1970, U.S. oil never recovered.

*We've embarked on the beginning of the last days of the age of oil,"*
*– Mike Bowlin, 1999, Chairman of ARCO and former Chairman of the Board of the American Petroleum Institute.*

That global oil output will eventually reach a peak and then decline is no longer a matter of debate; all major energy organizations have now embraced this view. What remains open for argument is precisely when this moment will arrive. Whatever the timing of this momentous event, it is apparent that the world faces a profound shift in the global availability of energy, as we move from a situation of relative abundance to one of relative scarcity.

Energy experts have long warned that global oil and gas supplies are not likely to be sufficiently expandable to meet anticipated demand. As far back as the mid-1990s, peak-oil theorists like Kenneth Deffeyes of Princeton University and Colin Campbell of the Association for the Study of Peak Oil (ASPO) insisted that the world was heading for a peak-oil moment and would soon face declining petroleum output. At first, most mainstream experts dismissed these claims as simplistic and erroneous, while government officials and representatives of the big oil companies derided them. Recently, however, a sea-change in elite opinion has been evident. First Matthew Simmons, the chairman of Simmons and Company International of Houston, America's leading energy-industry investment bank, and then David O'Reilly, CEO of Chevron, the country's second largest oil firm, broke ranks with their fellow oil magnates and embraced the peak-oil thesis. O'Reilly has been particularly outspoken, taking full-page ads in the New York Times and other papers to declare, "One thing is clear: the era of easy oil is over."

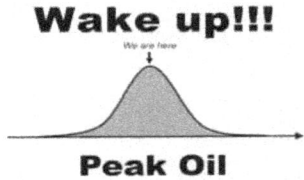

> *"You can produce motorcars to consume all the petroleum within the earth, and then you become no petrol. Then throw all these motorcars. Unless you find out some other energy. That you can do. You can make things topsy-turvied. But by your so-called scientific advancement, you cannot increase the supply."*
> ~ *Srila Prabhupada (Lecture on Srimad-Bhagavatam, Los Angeles,*

The exact moment of peak oil's arrival is not as important as the fact that world oil output will almost certainly fall short of global demand, given the fossil-fuel voraciousness of the older industrialized nations, especially the United States, and soaring demand from China, India, and other rapidly growing countries. The U.S. Department of Energy (DoE) projects global oil demand to grow by 35% between 2004 and 2025 -- from 82 million to 111 million barrels per day.

Much of the world's easy-to-acquire petroleum has already been extracted and significant portions of what remains can only be found in places that present significant drilling challenges like the hurricane-prone Gulf of Mexico or the iceberg-infested waters of the North Atlantic -- or in perennially conflict-ridden and sabotage-vulnerable areas of Africa, Central Asia, and the Middle East.

Compounding the global impact of America's extreme vulnerability is the highly problematic fact that the very same geological circumstances that have caused U.S. production to be halved since 1970, from 10 to 5 million barrels a day, are very much the same set of circumstances that apply to the entire planet as an oil producer. Oil is finite, production will peak and then inexonerably decline. 33 out of the top 48 producing nations like the U.S. have passed their peak in terms of production.

Although we cannot hope to foresee all the ways such forces will affect the global human community, the primary vectors of the permanent energy crisis can be identified and charted. Three such vectors, in particular, demand attention: a slowing in the growth of energy supplies at a time of accelerating worldwide demand; rising political instability provoked by geopolitical competition for those supplies; and mounting environmental woes produced by our continuing addiction to oil, natural gas, and coal. Each of these would be cause enough for worry, but it is their intersection that we need to

*"We don't have to run out of oil to start having severe problems with industrial civilization and its dependent systems. We only have to slip over the all-time production peak and begin a slide down the arc of steady depletion."*     *- James Howard Kunster*

fear above all.

Major oil finds (of over 500m barrels) peaked in 1964. In 2000, there were 13 such discoveries, in 2001 six, in 2002 two and in 2003 none. Three major new projects will come onstream in 2007 and three in 2008. For the following years, none have yet been scheduled.

One of the surprises in the oil world in 2004 was the success of a documentary on the perilous state of world energy. "The End of Suburbia." It has sold more than a million DVDs and has been aired on TV networks around the world.

With a global oil crisis looming like the Doomsday Rock, why do so few political leaders seem to care? Many experts refuse to take the problem seriously because it "falls outside the mind-set of market economics." Thanks to the triumph of global capitalism, the free-market model now reigns almost everywhere. The trouble is, its principles "tend to break down when applied to natural resources like oil." The result is both potentially catastrophic and all too human. Our high priests - the market economists - are blind to a reality that in their cosmology cannot exist.

David Fleming writes in the British magazine Prospect (Nov. 2000), If you're wondering why you never hear of peak oil from the oil companies and government, remember what happened in January 2004. Here is a BBC report:

### Shell shares dive as proven oil reserves cut

*Despite rising profits, investors have turned their back on Shell. Giant oil group Royal Dutch Shell has said it is trimming its figures for proven oil and gas reserves by 20%. Stunned investors promptly began a sell-off that knocked more than 7% off the Anglo-Dutch firm's share price in both London and Amsterdam. Shell said it does not expect the reassessment to have any impact on its financial results, as 90% of the reserves involved remain undeveloped. But analysts were unconvinced. Shares in fellow oil firm BP also fell 2%.*

*"Although there is growing awareness of the problem, there is also widespread ignorance and denial, even by people who should know better. Mankind has, it seems, an infinite capacity for denial."*

*Investors and oil analysts were startled, and puzzled, by the move.*
*"It was shocking, to say the least," the Agence France Presse news agency*
*quoted one oil analyst who did not wish to be named as saying. "They gave*
*no detailed explanation why this has happened."*
*"This reduces the value of the company by 10% using discounted cash*
*flows," said Richard Brackenhoff, an oil analyst for Kempen & Co.*

Eventually, the chairman was forced to resign. Barely had the shares begun to rise than another reserve cut in March knocked them down again.

You can imagine the effects on the stock market if the oil companies admitted that oil was going to decline every year from now on and never recover. One day they *will* have to admit it but no company (or chairman) wants to be the first.

Of course now the perception is rapidly changing. Chevron, an oil major in the world, displays the following message on its homepage (WillYouJoinUs.com)

*"Energy will be one of the defining issues of the century. One thing is clear: the era of easy oil is over. So let the discussion begin. How will we meet the energy needs of the entire world in this century and beyond?"*

Using the known amount of available oil and the present rate of consumption, how long would it be before all that oil is used up?

This is known as the R/P ratio in the oil business. It may surprise you to know that in the BP Statistical Review for 2005 (using data from 2004), the length of time is 40.5 years. So, any person under the age of about thirty or forty would be likely to have to face a world without any oil.

---

*"At present the heavily industrialized United States, with only 5% of the world's population, is using more than 40% of the world's energy output. But how long can this situation last? To catch up to the United States, the rest of the world is racing to industrialize, but the world's limited energy reserves make the end of the energy bonanza inevitable."*
*~ Balavanta dasa*

## 'Peak Oil' Enters Mainstream Debate

A few years ago only a handful of geologists and academics were considering such a possibility. But now it appears even governments are taking a serious look at the subject.

The question is occupying more and more minds around the world. It could happen soon. A French government report on the global oil industry forecasts a possible peak in world production as early as 2013.

The report 'The Oil Industry 2004' takes a long look at future production and supply issues. But perhaps what is most interesting about this Economics, Industry & Finance Ministry report, is that it actually mentions Peak Oil and a possible production plateau.

Even one year ago it was unheard of to find the subject mentioned amongst government ministries or financial institutions. Now banks such as Goldman Sachs, Caisse D'Epargne/Ixis, Simmons International and the Bank of Montreal have all broached the subject.

## No Escape From Scarcity

To make the energy picture grimmer, "spare" or "surge" capacity seems to be disappearing in the major oil-producing regions. At one time, key producers like Saudi Arabia retained an excess production capacity, allowing them to rapidly boost their output in times of potential energy crisis like the 1990-91 Gulf War. But Saudi Arabia, like the other big suppliers, is now producing at full tilt and so possesses zero capacity to increase output. In other words, any politically inspired (or sabotage related) cutoff in oil exports from countries like Russia or Iran will produce instant energy shock on a global scale and send oil prices soaring to, or through, that $200 a barrel barrier.

A chronic shortage of oil would be hard enough for the world community to cope with even if other sources of energy were in great

*"I will work for energy policies that recognize oil won't last forever."*
~Roscoe Bartlett

supply. But this is not the case. Natural gas -- the world's second leading source of energy -- is also at risk of future shortages. While there are still major deposits of gas in Russia and Iran (potentially the world's number one and two suppliers) waiting to be tapped, obstacles to their exploitation loom large. The United States is doing everything it can to prevent Iran from exporting its gas (for example, by strong-arming India into abandoning a proposed gas pipeline from Iran), while Moscow has actively

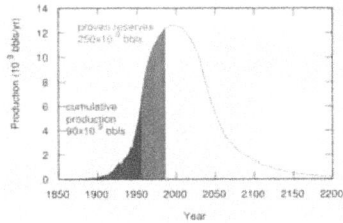

discouraged Europe from increasing its reliance on Russian gas through its recent cutoff of supplies to Ukraine and other worrisome actions.

## Sleepwalking Into The Future

American natural-gas production is declining at five percent a year, despite frantic new drilling, and with the potential of much steeper declines ahead. Because of the oil crises of the 1970s, the nuclear-plant disasters at Three Mile Island and Chernobyl and the acid-rain problem, the U.S. chose to make gas its first choice for electric-power generation. The result was that just about every power plant built after 1980 has to run on gas. Half the homes in America are heated with gas. To further complicate matters, gas isn't easy to import. In North America, it is distributed through a vast pipeline network. Gas imported from overseas would have to be compressed at minus-260 degrees Fahrenheit in pressurized tanker ships and unloaded (re-gasified) at special terminals, of which few exist in America. Moreover, the first attempts to site new terminals have met furious opposition because they are such ripe targets for terrorism.

> *"It took 500 million years to produce these hydrocarbon deposits and we are using them at a rate in excess of 1 million times their natural rate of production. On the time scale of centuries, we certainly cannot expect to continue using oil as freely and ubiquitously as we do today. Something is going to have to change."*

We are entering a historical period of potentially great instability, turbulence and hardship. Obviously, geopolitical maneuvering around the world's richest energy regions has already led to war and promises more international military conflict.

During campaign 2000, Bush told his countrymen that he had an energy plan that would reduce gas prices at the pumps and here we sit 7 years later, with the highest prices in history.

The Middle East will inevitably burn. It is a matter of either slowly, gradually, or in one almighty fire. That could mean anything from $200-300 per barrel, depending on the nature and the speed of developments.

We have become so dependent on those fuels, that there is no way we can sustain ourselves at this level of technology without them. Even something as basic as food becomes impossible to produce, process and transport without fuel.

The biggest news story of modern times rarely appears in the conventional news media, or it appears only in distorted forms. Ironically, the modern world is plagued by a lack of serious information. Today's news item is usually forgotten by tomorrow. The television viewer has the vague impression that something happened somewhere, but one could change channels all day without finding anything below the surface. But television is only the start of the enigma. What is most apparent is the larger problem that there is no leadership, no sense of organization, for dealing with peak-oil issues.

Politicians whose careers span an average of 5 years, clearly don't see it as their problem. Let the next guy deal with it if he has to.

Although there is growing awareness of the problem, there is also widespread ignorance and denial, even by people who should know better.

Mankind has, it seems, an infinite capacity for denial but still the message is gradually sinking in: the term "Peak Oil" appears in the

*The oil crisis gets louder – listen to it, talk about it, prepare for it – it is out there, the tide is rising and rushing towards us.*
*-Dan Smith*

press with increasing frequency.

Number of producing oil wells is declining. Each year "World Oil Organization" produces a review of the previous years. In 1997 they had 918,896 producing oil wells world wide. This declined to 902,103 in 1998, and to 879,8888 in 1999. It recovered a bit in 2000 rising to 884,843. The trend is clearly downward.

## Gulf of Mexico

### Platform Removals to Exceed Platform Installations

When an oil field is found, an oil company will install a platform which will house the production equipment. At the end of the field's life, the platform must be removed. In all but one year over the past 50 years, more platforms have been installed than removed. But starting this year or next, the Gulf will enter a period in which more platforms are removed than are installed. The MMS (Minerals Management Service) predicts that over the next 25 years there will be an average of 142 installations per year and 186 removals per year.

The world has experienced severe energy crises before: the 1973-74 "oil shock" with its mile-long gas lines; the 1979-80 crisis following the fall of the Shah of Iran; the 2000-01 electricity blackouts in major cities among others. But the crisis taking shape today has a new look to it. First of all, it is likely to last for decades, not just months or a handful of years; second, it will engulf the entire planet, not just a few countries; and finally, it will do more than just cripple the global economy -- its political, military, and environmental effects will be equally severe.

*Chevron, a major oil company in the world, its homepage (WillYouJoinUs) reads:*
*"Energy will be one of the defining issues of the century. One thing is clear: the era of easy oil is over. So let the discussion begin. How will we meet the energy needs of the entire world in this century and beyond?"*

# BP Says Global Oil Reserves Growth Stalled In 2004

*Reuters.co.uk*
*By Tom Bergin, European Oil and Gas Correspondent*

LONDON (Reuters) - Growth in the world's oil and gas reserves stalled last year, a report from oil giant BP showed on Tuesday, bucking a trend that has historically seen new discoveries more than match production.

The BP Statistical Review of World Energy, compiled from official government figures, will reinforce concerns about the ability of global oil supplies to match surging consumption, which grew 3.4 percent in 2004.

The world had 1,188.6 billion barrels of oil reserves at the end of 2004, compared to 1,188.3 billion at the end of 2003, BP, the world's second largest oil firm by market capitalisation, said.

The 0.02 percent growth rate was the lowest since 1990 and compares with a 10-year average above 1.5 percent per annum.

Last year's almost imperceptible rise in oil reserves came despite high prices, which normally help by encouraging new exploration and by making previously uneconomic resources commercial.

Gas fared only slightly better with reserves growing 0.18 percent, but this was the lowest growth rate in over 20 years, and well below

*"The world has never faced a problem like Peak Oil. Without massive mitigation more than a decade before the fact, the problem will be pervasive and will not be temporary. Previous energy transitions (wood to coal and coal to oil) were gradual and evolutionary; oil peaking will be abrupt and revolutionary".*
*~Neal Brandvik*

the 10-year average of more than 2 percent each year.

The figures contrast with BP's view, regularly voiced by Chief Executive John Browne, that the world is not facing a supply crunch.

However, the data echoes the oil majors' own difficulties in finding oil. Last year, the biggest international firms replaced around 70 percent of the oil and gas they pumped with new finds, analysts said.

Even BP, one of the better explorers in the industry, failed to achieve the 100 percent reserve replacement ratio that shows a firm's resource base is not shrinking.

The report also points to another worrying trend for the oil majors. The gap between their anaemic reserve replacement ratio and an effective 100 percent ratio globally supports investors' fears that the biggest oil companies will lose market share.

Analysts have predicted that firms like BP and U.S. rival Exxon Mobil will become increasingly constrained in finding new exploration opportunities in the future because the biggest hydrocarbon reserves look set to be controlled by state-owned oil and gas companies in Russia, Venezuela and the Gulf states.

BP cautioned that pundits have been predicting the imminent depletion of reserves for a century and added that since different governments use different methodologies to calculate proved reserves, it is hard to draw inferences from its review, which is published annually.

*"This motorcar civilization will be finished within another hundred years. It has begun, say, for the last hundred years, and after a hundred years, when... The scientists say the petroleum will be finished within fifty years or like that, so, say hundred years, this motorcar will be finished."*
~*Srila Prabhupada (Lecture on Srimad-Bhagavatam, Los Angeles, August 18, 1972)*

# Key Oil Figures Were Distorted By US Pressure, Says Whistleblower

## Watchdog's Estimates of Reserves Inflated Says Top Official

*By Terry Macalister*
*guardian.co.uk, Monday 9 November 2009 21.30 GMT*

The world is much closer to running out of oil than official estimates admit, according to a whistleblower at the International Energy Agency who claims it has been deliberately underplaying a looming shortage for fear of triggering panic buying.

The senior official claims the US has played an influential role in encouraging the watchdog to underplay the rate of decline from existing oil fields while overplaying the chances of finding new reserves.

The allegations raise serious questions about the accuracy of the organisation's latest World Energy Outlook on oil demand and supply to be published tomorrow – which is used by the British and many other governments to help guide their wider energy and climate change policies.

'There's suspicion the IEA has been influenced by the US' Link to this audio

In particular they question the prediction in the last World Economic Outlook, believed to be repeated again this year, that oil production can be raised from its current level of 83m barrels a day to 105m barrels. External critics have frequently argued that this cannot be substantiated by firm evidence and say the world has already passed its peak in oil production.

Now the "peak oil" theory is gaining support at the heart of the global energy establishment. "The IEA in 2005 was predicting oil supplies could rise as high as 120m barrels a day by 2030 although it was forced to reduce this gradually to 116m and then 105m last year," said the IEA source, who was unwilling to be identified for fear of reprisals inside the industry. "The 120m figure always was nonsense but even today's number is much higher than can be justified and the IEA knows this.

"Many inside the organisation believe that maintaining oil supplies at even 90m to 95m barrels a day would be impossible but there are fears that panic could spread on the financial markets if the figures were brought down further. And the Americans fear the end of oil supremacy because it would threaten their power over access to oil resources," he added.

A second senior IEA source, who has now left but was also unwilling to give his name, said a key rule at the organisation was that it was "imperative not to anger the Americans" but the fact was that there was not as much oil in the world as had been admitted. "We have [already] entered the 'peak oil' zone. I think that the situation is really bad," he added.

"That's all folks!"

The IEA acknowledges the importance of its own figures, boasting on its website: "The IEA governments and industry from all across the globe have come to rely on the World Energy Outlook to provide a consistent basis on which they can formulate policies and design business plans."

The British government, among others, always uses the IEA statistics rather than any of its own to argue that there is little threat to long-term oil supplies.

The IEA said tonight that peak oil critics had often wrongly questioned the accuracy of its figures. A spokesman said it was unable to comment ahead of the 2009 report being released tomorrow.

John Hemming, the MP who chairs the all-party parliamentary group on peak oil and gas, said the revelations confirmed his suspicions that the IEA underplayed how quickly the world was running out and this had profound implications for British government energy policy.

He said he had also been contacted by some IEA officials unhappy with its lack of independent scepticism over predictions. "Reliance on IEA reports has been used to justify claims that oil and gas supplies will not peak before 2030. It is clear now that this will not be the case and the IEA figures cannot be relied on," said Hemming.

"This all gives an importance to the Copenhagen [climate change] talks and an urgent need for the UK to move faster towards a more sustainable [lower carbon] economy if it is to avoid severe economic dislocation," he added.

The IEA was established in 1974 after the oil crisis in an attempt to try to safeguard energy supplies to the west. The World Energy Outlook is produced annually under the control of the IEA's chief economist, Fatih Birol, who has defended the projections from earlier outside attack. Peak oil critics have often questioned the IEA figures.

"WE'D LIKE A GPS DEVICE THAT WOULD DIRECT US TO AFFORDABLE GAS."

But now IEA sources who have contacted the Guardian say that Birol has increasingly been facing questions about the figures inside the organisation.

Matt Simmons, a respected oil industry expert, has long questioned the decline rates and oil statistics provided by Saudi Arabia on its own fields. He has raised questions about whether peak oil is much closer than many have accepted.

A report by the UK Energy Research Centre (UKERC) last month said worldwide production of conventionally extracted oil could "peak" and go into terminal decline before 2020 – but that the government was not facing up to the risk. Steve Sorrell, chief author of the report, said forecasts suggesting oil production will not peak before 2030 were "at best optimistic and at worst implausible".

But as far back as 2004 there have been people making similar warnings. Colin Campbell, a former executive with Total of France told a conference: "If the real [oil reserve] figures were to come out there would be panic on the stock markets … in the end that would suit no one."

(*guardian.co.uk*)

# The End Of Fossil Fuel

*forbes.com*
*By Chris Nelder, 07.24.09, 03:00 PM EDT*

*Prepare for a radically different lifestyle as global crude oil production peaks and begins to decline.*

*You will never see cheap gasoline again. You will probably never see cheap energy again. Oil, natural gas and coal are set to peak and go into decline within the next decade, and no technology can change that.*

Peaking is a simple concept. We generally exploit natural resources in a bell-shaped curve, with the rate of extraction increasing over time until we reach a peak and then gradually slowing down until we stop using them.

Peak oil is not about "running out of oil"; it's about reaching the peak rate of oil production. It's not the size of the tank that matters, but the size of the tap.

The peak is usually reached when resources become too difficult to extract, or too expensive, or they are replaced by something cheaper, better or more plentiful. Unfortunately, we have no substitutes for oil that are cheaper or better.

According to the best available data, we are now at the peak rate of oil production. After over a century of continual growth, global conventional crude oil production topped out in 2005 at just over 74 million barrels per day (mbpd) and has remained at that level ever since.

The additional "oil" that brings the oft-cited world total to 84 mbpd today (down from 87 mbpd last year; according to U.S. government data) isn't conventional crude, but, rather, unconventional hydrocarbons, including natural gas liquids, "extra heavy" oil, synthetic oil made from Canadian tar sands, refinery gains, liquids produced from the conversion of coal and natural gas, and biofuels.

Oil production is expected to go into terminal decline around 2014. The principal reason is that the largest and most productive fields are becoming depleted while new discoveries have been progressively smaller and of lesser quality. Discovery of new oil peaked over 40 years ago and has been declining ever since despite furious drilling and unprecedentedly high prices.

When it begins to decline, rate of crude production is projected to fall at 5%, or over four mbpd, per year--roughly equivalent to losing the entire production of Latin America or Europe every year. The decline rate will likely accelerate to over 10% per year by 2030.

The Paris-based International Energy Agency estimates that the world would need to add the equivalent of six new Saudi Arabias by 2030 in order to meet declining production and growing demand. Obviously, there aren't another six Saudi Arabias waiting to be discovered, and unconventional liquid fuels simply cannot fill such a yawning gap.

Natural gas is likewise expected to peak some time around 2010-2020, and coal around 2020-2030. Oil, natural gas and coal together provide 86% of the world's primary energy.

By the end of this century, nearly all of the economically recoverable fossil fuels will be gone. From now until then, what remains will be rationed by price. There will be shortages.

Renewable energy--solar, wind, geothermal--currently makes up less than 2% of the world's primary energy supply, and although growing very rapidly, it is not on course to fill the fossil fuel gap, either.

As fossil fuels peak and then decline, the world's economies will be forced for the first time to live within a shrinking, not expanding, energy budget. They will adapt to this new reality by repeating the cycle we saw over the last 18 months: commodity price spikes, leading to economic destruction, leading to supply destruction, leading back

to price spikes. Only in recessionary periods, like now, will there be excess supply.

How this will affect the global economy, and our lifestyles, cannot be overstated. Former chief economist for Canadian Imperial Bank of Commerce World Markets, Jeff Rubin, and oil investment banker Matthew Simmons have concluded that it means no less than the end of globalization.

bbc.co.uk — Home | TV | Radio | Talk | Where I Live | A-Z
UK version ● International version | About the versions

BBC NEWS — ▶ WATCH LIVE BBC News 24

News Front Page — Last Updated: Tuesday, 5 September 2006, 10:19 GMT 11:19 UK
World — ✉ E-mail this to a friend — 🖨 Printable version

**Has oil production finally peaked?**

With the demand for energy increasing, are we edging closer to "peak oil" - and what can be done about it?

Stephen Leeb, founder of Leeb Capital Management Group and a long-time analyst on Wall Street, thinks so.

"We have a president that says we're addicted to oil, but doesn't say that we don't have enough oil to satisfy our addiction," he says. "He really hasn't alerted us to the fact that it's a true crisis."

Americans, who constitute 4% of the world population but consume 25% of its energy, will have radically different lifestyles. Production of everything will have to be re-localized. Instead of our food traveling an average 1,500 miles before it reaches us, it will have to come from nearby and use organic methods instead of requiring 10 calories of fossil fuel inputs for every calorie of food we eat.

*I was in New York in the 30's. I had a box seat at the depression. I can assure you it was a very educational experience. We shut the country down because of monetary reasons. We had manpower and abundant raw materials. Yet we shut the country down. We're doing the same kind of thing now but with a different material outlook. We are not in the position we were in 1929–30 with regard to the future. Then the physical system was ready to roll. This time it's not. We are in a crisis in the evolution of human society. It's unique to both human and geologic history. It has never happened before and it can't possibly happen again. You can only use oil once. You can only use metals once. Soon all the oil is going to be burned and all the metals mined and scattered.*

*~M. King Hubbert, 1983*

Rather than shipping ore to China and shipping it back to the U.S. as steel, we'll need to revive our domestic steel industry. "Bedroom communities" will die and ideally be reborn as fully functional independent communities. It means the end of long commutes.

The coming energy shortage is the most serious crisis the world has ever faced, but it could have a very positive outcome. In theory, the Earth's wind, solar, geothermal and marine resources could each provide more than the total energy the world consumes every day, if we had the ability to harvest them.

As fossil fuel prices rise, the price of renewably generated electricity will continue to fall. If we are wise and lucky, we will rapidly improve the efficiency of our built environment, deploy renewable capacity and convert to an all-electric infrastructure that runs on it. Fortunately, political momentum is now leaning strongly in this direction.

If we move fast to re-localize production and proceed with the renewable revolution, we could end the 21st century with a largely carbon-free economy, putting an end to climate change and averting resource wars. We would have healthier food and a safer, more resilient and equitable world.

*(Chris Nelder is the author of Profit from the Peak--The End of Oil and the Greatest Investment Event of the Century and the coauthor of Investing in Renewable Energy.)*

# 'Peak oil' debate is no longer on hold
*BusinessDay, South Africa, 2009/07/21*

Put a group of oil experts under one roof for a while and their discussion is likely to drift to the subject of peak oil — a point in time when maximum oil production is reached, after which it goes into permanent decline.

The advent of peak oil has long been brushed aside by some because it seems like a far-fetched, if not a ridiculous, idea concocted by alarmists. This is despite deafening cries that it is a real and serious threat.

Even among those who agree that it will happen, views differ sharply on the date . Some, like author David Strahan, say it could be as soon as 2017.

Recent data show that the debate can no longer be dismissed as a figment of the imagination among peak oil "enthusiasts".

According to the Washington, US-based Worldwatch Institute, oil production is in decline in 33 of the 48 largest oil-producing countries. The research organisation says most of these countries are past their oil production peaks. Iran peaked in 1974, Nigeria in 1979, Venezuela in 1970 and Mexico in 2004.

Saudi Arabia, the world's largest oil exporter, is expected to reach its peak in 2014, while in Iraq this is estimated in 2018.

Last year's study by professional services group Ernst & Young showed that in the period between 2003-07, oil production in the US remained flat at about 1,2-million barrels a day.

Oil companies had difficulty in finding investment and production opportunities, say Ernst & Young.

But not everyone is convinced about peak oil. BP chief economist Christof Rühl says the argument for peak oil is baseless. "Peak oil has been predicted for 150 years. It has never happened, and will stay this way," Rühl has reportedly said. He says oil is about price and not about availability.

Economist Tony Twine of consultants Econometrix echoes the view that price is everything.

"All energy — gas, oil and coal — is exploitable at a given price. If the price falls below a particular price it becomes worthless to produce. That is why I say many of the peak oil arguments are not well based.

"They all assume an oil price at 30, 60 or 200 a barrel," he says. What is known as "oil availability" differs at different oil prices, Twine says.

"The projections that are being made about peak oil are sensible in particular contexts. But whether they are universally true is another matter," he says.

Even in 30-50 years' time, if oil demand is greater than supply, oil prices will rise "and currently unexploitable deposits will become viable to exploit", Twine says. Oil wells now considered marginal will become profitable .

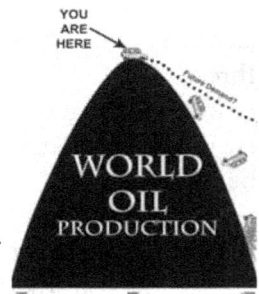

Twine says there is a tendency to look at oil in terms of its energy content. "But there is a range of products that come out of a barrel of oil — from fertiliser to solvents that end up in paints, washing powder and synthetic fibres. Almost anything that you can see and feel has a little bit of oil in it.

"As oil becomes scarce and more expensive, its use as a source of energy will diminish. But its use as a feedstock for the chemicals industry will take longer to disappear," Twine says.

Richard Worthington, climate change programme manager for the World Wildlife Fund in SA, says the advent of peak oil should influence how hydrocarbons are used. "It highlights the need for greater efficiency," he says. Climate change considerations have superseded peak oil discussions.

Worthington says fears of peak oil should not be the main driver of the move away from fossil- based energy sources. At some stage fossils will be depleted, he says. "Now there is talk of peak oil, then it will be peak energy and then peak coal," he says.

Indeed, depletion of gas and coal reserves is a double whammy. National oil and gas company PetroSA's Mossel Bay gas-to- liquids refinery is set to run out of natural gas by 2011.

The offshore fields south of Mossel Bay will not be able to keep up the supply of 36000 barrels a day the refinery needs.

The dwindling gas reserves are to be expected, says Twine.

"Gas and oil fields in SA and Mozambique have always been known to be constrained in terms of reserves. They have always been marginal in terms of big investment spending," Twine says.

However, he believes that the Mozambique gas fields will have a longer life span and are likely to fuel petrochemicals group Sasol for a longer time. Sasol's synfuels plant in Secunda gets natural gas from Mozambique through an 865km-long pipeline.

*(© BusinessDay, SA)*

*Initially it will be denied. There will be much lying and obfuscation. Then prices will rise and demand will fall. The rich will outbid the poor for available supplies. The system will initially appear to rebalance. The dash for gas will become more frenzied. People will realize nuclear power stations take up to ten years to build. People will also realize wind, waves, solar and other renewables are all pretty marginal and take a lot of energy to construct. There will be a dash for more fuel-efficient vehicles and equipment. The poor will not be able to afford the investment or the fuel. Exploration and exploitation of oil and gas will become completely frenzied. More and more countries will decide to reserve oil and later gas supplies for their own people. Air quality will be ignored as coal production and consumption expand once more. Once the decline really gets under way, liquids production will fall relentlessly by five percent per year. Energy prices will rise remorselessly. Inflation will become endemic. Resource conflicts will break out.*
*~Colin Campbell, March 2002*

# Oil And India

*By Vandana Shiva*

Oil is a non-renewable resource. We have always known that yet the world has been behaving as if oil is in endless supply. And we in India who have lived in a biodiversity and biomass energy economy are rushing into oil addiction precisely when the global oil supply is running low and prices are running high.

The Association for the Study of Peak Oil (ASPO), an umbrella organization of oil experts, mainly geologists who helped find oil fields are now warning us that there are only a trillion barrels or less of oil left, and the supply will peak within this decade. "Peak Oil", or the topping point, is the highest amount that can ever be pumped. Beyond "peak oil", there will be an overall decline in production and an increase in oil prices. Oil that costs $5 per barrel to extract could become $ 100 per barrel when confidence in supply erodes and demand increases, and there is recognition that we are in a world of shrinking oil supplies, not growing supplies.

Why are we as a country tying our future to a resource that must shrink and become more costly? As we build more superhighways and mega cities, destroying the decentralized fabric of our socio-economic organization, we need to ask how long will this last?

There is another reason to stop this frenzy of oil addiction, and that is climate change, or more accurately, climate chaos. Climate change is caused by fossil fuel emissions, and stabilizing carbon dioxide emissions is an ecological imperative. This is why the Kyoto Protocol to the climate change convention was signed. The insurance industry, which takes over $ 2 trillion in annual premiums, and is bigger than the oil industry, is now a major player in addressing climate change since they have to pay billions out in insurance as cities flood, cyclones such as Katrina uproot entire communities and heat waves kill.

The costs of climate change to the people of India are extremely high. The 1999 Orissa super cyclone and the Bombay floods of July 2006 are just two better-known extreme events linked to a changing climate.

This winter, we had no rains during the wheat season, and heavy downpours during the wheat harvest. Heavy rains before the monsoon in the catchments of the Ganga and Yamuna destroyed crops so that farmers did not even have seeds to sow. And in Sikkim, heavy rains led to landslides, which disrupted Gangtok's water supply. I was in Sikkim during the crisis and we lived on one bucket a day.

The fossil fuel economy is based on two illusions - one, that we can keep up our oil addiction, and two, that substituting renewable energy with fossil fuel has only benefits, no costs. Climate change is very high cost of an economy based on oil. We are starting to eat oil and drink oil. Oil is at the heart of industrial food production and processing, and long distance food transport. The wheat, India is importing is not just bringing weeds, pests and pesticides. It is also carrying thousands of "food miles". Imagine a Tsunami or cyclone if our food supplies become dependent on wheat from U.S and Australia. And imagine the cost of wheat as oil prices rise, and wheat embodies more oil than nutrition.

We are also drinking oil, not water. When Coca Cola and Pepsi pump 1.5 to 2 million gallons a day to fill their soft drink and water

*"For a successful technology, reality must take precedence over public relations, for Nature cannot be fooled."*
*-Richard Feynman*

bottles, and transport them to the remotest part of India, water embodies oil both in its extraction and transport. It is increasingly impossible to find clean water in our wells and springs. But Aqua Fina and Kinley has reached every village, selling water which has become oil, packaged in a plastic bottle made from oil.

While the political parties protest against the hike in oil prices, society also needs to start taking a long-term view of the ecological, economic and social costs of our growing oil addition. We need to start addressing strategic issues of real and sustainable energy security in the context of peak oil, the end of cheap oil, and the climate chaos that the era of cheap oil has left as an environmental burden on the planet.

*(Dr. Vandana Shiva is a physicist, ecologist, activist, editor, and author of many books.)*

# Future Demand Will Outstrip Supply And OPEC Can't Help

*By A. F. Alhajji*

Either EIA &IEA projections are wrong, or a crisis appears to be imminent. The World Energy Outlook 2000, compiled by the U.S. Energy Information Administration (EIA), and the International Energy Outlook 2001, authored by the International Energy Agence (IEA), indicate that oil production in the Arabian Gulf States must almost double by year 2020 to meet rising world demand.

The EIA outlooks states, 'The reference case projection implies aggressive efforts by investment capital, to implement a wide range of production capacity expansion. However, the combination of potential profitability and the threat of competition from non-OPEC supplies argues for the pursuit of an aggressive expansion strategy.'

"The reference case requires Gulf states to increase their oil production 80% by 2020. This means adding about 13 million bopd

> *Petrol..., you have created another problem. Already there are problems. You have created, by so-called civilization, petrol problem. Before these motorcars, the people were living very happily. They were transporting. But there was no such civilization that for your earning livelihood you have to go hundred miles away from your home to work there. Therefore you require vehicle. Then you require petrol. Then you require so many nice road. So many things will be. But formerly, it was village. They will take it, "This is primitive." But remaining primitive, you were more happy than becoming so-called civilized, creating so many problems.*
> *~ Srila Prabhupada (Lecture on Srimad-Bhagavatam, Hawaii, January 17, 1974)*

by 2020--with Saudi Arabia increasing its capacity by more than 7 million bopd, to about 17 million bopd. This seems highly unrealistic. So, either EIA & IEA or a supply crisis is coming, because Saudi Arabia and its neighbors cannot increase production by 80% for many technical, financial and political reasons."

Mistaken projections. Two recent studies by prominent oil market experts Guy Caruso (Center for Strategic and International Studies, Washington) and Prof. Deromt Gately (New York University) show that such EIA and IEA projections are wrong. In his study, 'How likely is the consensus projection of oil production doubling in the Persian Gulf?', Gately states, 'Such projections are not based on behavioral analysis of Gulf countries' decisions. They are merely the calculated residual demand for OPEC oil, the difference between projected world oil demand and non-OPEC oil supply.'

( A. F. Alhajji, "Will Gulf States Live Up to EIA and IEA Projections?"" World Oil, June 2001)

*People are always saying the world will end and it never does. Maybe it won't this time, either. But, frankly, it's not looking good.*
*~Norman Church*

## A World Following In The Footsteps of
## United States of Amnesia
*By Jeff Berg*

For starters the U.S. is the largest consumer of oil in the world by far, it's not even close, its been that way for a hundred years, and still this isn't common knowledge. Today the U.S. is consuming 20 million barrels a day, 14 million of those barrels imported. India by contrast is sixth in the world in terms of oil consumption consuming about 2.5 Mb/d (million barrels per day) or just about 1/8 of the U.S. amount. Oh and by the by while they do consume but 1/8 of what the U.S. does on a daily basis they have four times as many people meaning of course that they are consuming but 1/36th (3%) of the U.S. amount on a per capita basis.

No single country in the world comes close to the U.S. numbers, the two next greatest consumers of oil, China and Japan each consume about 1/4 of what the U.S. does. In fact in order to equal U.S. consumption you have to add China, Japan, Germany, Russia and India together. The population of these countries being about 2.75 billion to the U.S. 300 million (9.25 times greater.)

Is anyone going my way?

The U.S. leadership knows its extreme vulnerability and has known it for many decades. So well ingrained is this fact among America's Presidents, its intelligence agencies, the State Department and the Pentagon that Jimmy Carter announced to the world that nothing would stand in the way of the U.S. and its interests in the Middle

East region. (Now referred to as 'The Carter Doctrine') Today we are seeing just exactly how seriously these words were meant.

The U.S. is a massive and insane energy hog recklessly dependent on the rest of the world for 2/3 of its liquid fuel needs. Its a dangerous weakness and it is spending an extremely precious and finite resource like a "drunken sailor on leave."

Our world leaders' response to Peak Oil

As long as this insanity is confined to this 300 million, it is probably manageable. But suddenly, the third World is waking up and trying to live the American dream and that is a very dangerous turn for humanity.

*As we all know, no crude oil refineries have been built in the United States since 1976. During that time, close to 100 ethanol refineries have been built.*
*~John Shimkus*

# And There Goes Mexico

*By David Brown*

AAPG Explorer had an article on doing business in Mexico. It had a fascinating, and fearful statement by a major player in the Mexican oil industry. Luis Ramirez Corzo, the Director-general of exploration and production for Petroleros Mexiconos (Pemex) looked at his country's condition and said:

"Production from Cantarell will begin to decline in 2006, and the drop-off will be brutal—14 percent a year, according to Ramirez."
"When you put the numbers together, it looks like the production in Mexico is going to decline over the next few years, and the decline is going to be quite steep,' he said.
"This situation has long set off alarms inside Pemex. But the rest of Mexico doesn't see the scope of the coming challenge, according to Shaw.
'I don't think the public in general understands that they're about to hit a brick wall,' he observed."
*(David Brown, "Politics Cloud Mexico's Promises," AAPG Explorer, Oct. 2004, p. 16)*

*"By increasing every year new motorcars, I am creating another problem. If there is no petrol, then the whole business will be spoiled." That they do not know. And because they do not know, they are called asses, mudha.*
*~Srila Prabhupada (Lecture on Srimad–Bhagavatam, Los Angeles, January 2, 1974)*

# Trouble In The World's Largest Oil Field - Ghawar

There are four oil fields in the world which produce over one million barrels per day. Ghawar, which produces 4.5 million barrels per day, Cantarell in Mexico, which produces nearly 2 million barrels per day, Burgan in Kuwait which produces 1.7 million barrels per day and Da Qing in China which produces 1 million barrels per day. Ghawar is, therefore, extremely important to the world's economy and well being. Today the world produces 82.5 million barrels per day which means that Ghawar produces 5.5 percent of the world's daily production. Should it decline, there would be major problems. As Ghawar goes, so goes Saudi Arabia.

The field was brought on line in 1951. By 1981 it was producing 5.7 million barrels per day. Its production was restricted during the 1980s but by 1996 with the addition of two other areas in the southern area of Ghawar, the production went back up above 5 million per day. In 2001 it was producing around 4.5 million barrels per day. There have been 3400 wells drilled into this reservoir.

"The big risk in Saudi Arabia is that Ghawar's rate of decline increases to an alarming point," says Ali Morteza Samsam Bakhtiari, a senior official with the National Iranian Oil Company. "That will set bells ringing all over the oil world because Ghawar underpins Saudi output and Saudi undergirds worldwide production."

*(Jeff Gerth, "Forecast of Rising Oil Demand Challenges Tired Saudi Fields," February 24, 2004 New York Times )*

## But This Is What Is Happening

"Saudi oilmen are usually a taciturn bunch, guarding their data like state secrets. But this was post September 11 and Riyadh was wooing western journalists and trying to restore the Saudis' image as dependable, long-term suppliers of energy—not suicidal fanatics nor terrorist financiers. And it was working. 'At Ghawar,' they have to inject water into the field to force the oil out,' by contrast, he continued, Shayba's oil contained only trace amounts of water. At Ghawar, the engineer said, the 'water cut' was 30%."

"Ghawar's water injections were hardly news, but a 30% water cut, if true, was startling. Most new oilfields produce almost pure oil or oil mixed with natural gas—with little water. Over time, however, as the oil is drawn out, operators must replace it with water to keep the oil flowing —until eventually what flows is almost pure water and the field is no longer worth operating." "Ghawar will not run dry overnight, but the beginning of the end of its oil is in sight."

*(Paul Roberts, "New Tyrants for Old as the Oil Starts to Run Out, " Sunday Times (News Review), May 16, 2004, p. 8)*

"Saudi Aramco is injecting a staggering 7 million barrels of sea water per day back into Ghawar, the world's largest oilfield, in order to prop up pressure. It accounts for 30% of Saudi oil reserves and up to 70% of daily output."

*("Doubts grow about Saudi As Global Swing Producer," Aberdeen Press & Journal Energy, April 5, 2004, p. 15)*

But several people are becoming concerned about the ability of the Saudi's to maintain production. Here is a tidbit from the Aberdeen Scotland Newspaper of a few weeks ago.

"It seems a growing number of analysts are falling into line with

> *"Now, because we have got big, big motorcars, we have to go thirty miles to find out a doctor. So the other inconveniences are also increased. Now we have to find out petrol and flatter the Arabians, "Give me petrol."*
> *~Srila Prabhupada (Lecture on Bhagavad-gita, London, March 11, 1975)*

the Simmons & Company International view that Saudi Arabia may be running out of steam and may not be able to perform the role of global swing producer for many more years, despite being credited with oil reserves in the order of 260 billion barrels. The Centre for Global Energy Studies hinted at the beginning of the year that the kingdom appeared to be heading for difficulties. Now one of its analysts has said that having reserves does not equate to production capacity. Citing the Haradh field, he said it required 500,000 barrels per day of water injection to get out 300,000 bpd of oil. Moreover the problem is even more serious in the Khurais field."

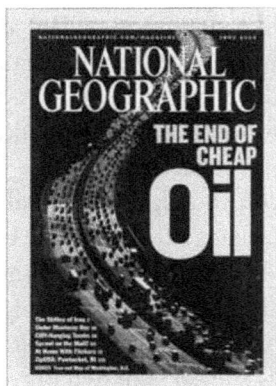

*(Doubts grow about Saudi As Global Swing Producer," Aberdeen Press & Journal Energy, April 5, 2004, p. 15)*

What is the future of Ghawar and Saudi production? It is not good.

"All production comes from 'very old fields', with no major exploration success since the 1960s, and almost every field has high and rising watercut.

*("Doubts grow about Saudi As Global Swing Producer," Aberdeen Press & Journal Energy, April 5, 2004, p. 15)*

As Ghawar goes, so goes the world.

*Any group of beings (human or nonhuman, plant or animal) who take more from their surroundings than they give back will, obviously, deplete their surroundings, after which they will either have to move, or their population will crash. ~ Bill Harding*

# The Pentagon V. Peak Oil

## How Wars of the Future May Be Fought Just to Run the Machines That Fight Them

*By Michael T. Klare*

Sixteen gallons of oil. That's how much the average American soldier in Iraq and Afghanistan consumes on a daily basis — either directly, through the use of Humvees, tanks, trucks, and helicopters, or indirectly, by calling in air strikes. Multiply this figure by 162,000 soldiers in Iraq, 24,000 in Afghanistan, and 30,000 in the surrounding region (including sailors aboard U.S. warships in the Persian Gulf) and you arrive at approximately 3.5 million gallons of oil: the daily petroleum tab for U.S. combat operations in the Middle East war zone.

Multiply that daily tab by 365 and you get 1.3 billion gallons: the estimated annual oil expenditure for U.S. combat operations in Southwest Asia. That's greater than the total annual oil usage of Bangladesh, population 150 million — and yet it's a gross underestimate of the Pentagon's wartime consumption.

Such numbers cannot do full justice to the extraordinary gas-guzzling expense of the wars in Iraq and Afghanistan. After all, for every soldier stationed "in theater," there are two more in transit, in training, or otherwise in line for eventual deployment to the war zone — soldiers who also consume enormous amounts of oil, even if less than their compatriots overseas. Moreover, to sustain an "expeditionary" army located halfway around the world, the Department of Defense must move millions of tons of arms, ammunition, food, fuel, and equipment every year by plane or ship, consuming additional tanker-loads of petroleum. Add this to the tally and the Pentagon's war-related oil budget jumps appreciably, though exactly how much we have no real way of knowing.

And foreign wars, sad to say, account for but a small fraction of the Pentagon's total petroleum consumption. Possessing the world's largest fleet of modern aircraft, helicopters, ships, tanks, armored vehicles, and support systems — virtually all powered by oil — the Department of Defense (DoD) is, in fact, the world's leading consumer of petroleum. It can be difficult to obtain precise details on the DoD's daily oil hit, but an April 2007 report by a defense contractor, LMI Government Consulting, suggests that the Pentagon might consume as much as 340,000 barrels (14 million gallons) every day. This is greater than the total national consumption of Sweden or Switzerland.

## Not Guns V. Butter, But Guns V. Oil

For anyone who drives a motor vehicle these days, this has ominous implications. With the price of gasoline now 75 cents to a dollar more than it was just six months ago, it's obvious that the Pentagon is facing a potentially serious budgetary crunch. Just like any ordinary American family, the DoD (department of defense) has to make some hard choices: It can use its normal amount of petroleum and pay more at the Pentagon's equivalent of the pump, while cutting back on other basic expenses; or it can cut back on its gas use in order to protect favored weapons systems under development. Of course, the DoD has a third option: It can go before Congress and plead for yet another supplemental budget hike, but this is sure to provoke renewed calls for a timetable for an American troop withdrawal from Iraq, and so is an unlikely prospect at this time.

*"Once the game is over, the king and the pawn go back in the same box."*
*~Italian Proverb*

Nor is this destined to prove a temporary issue. As recently as two years ago, the U.S. Department of Energy (DoE) was confidently predicting that the price of crude oil would hover in the $30 per barrel range for another quarter century or so, leading to gasoline prices of about $2 per gallon. But then came Hurricane Katrina, the crisis in Iran, the insurgency in southern Nigeria, and a host of other problems that tightened the oil market, prompting the DoE to raise its long-range price projection into the $50 per barrel range. This is the amount that figures in many current governmental budgetary forecasts — including, presumably, those of the Department of Defense. But just how realistic is this? The price of a barrel of crude oil today is hovering in the $66 range. Many energy analysts now say that a price range of $150-$200 per barrel (or possibly even significantly more) is far more likely to be our fate for the foreseeable future.

A price rise of this magnitude, when translated into the cost of gasoline, aviation fuel, diesel fuel, home-heating oil, and petrochemicals will play havoc with the budgets of families, farms, businesses, and local governments. Sooner or later, it will force people to make profound changes in their daily lives — as benign as purchasing a hybrid vehicle in place of an SUV or as painful as cutting back on home heating or health care simply to make an unavoidable drive to work. It will have an equally severe affect on the Pentagon budget. As the world's number one consumer of petroleum products, the DoD will obviously be disproportionately affected by a doubling in the price of crude oil. If it can't turn to Congress for redress, it will

*"The purpose is that you grow some castor seed, press it, get oil, put in any pot, and one wick, the light is there. So even understanding that you have improved the lighting system, but that is not the only necessity of my life. But to improve from the castor seed lamp, castor oil lamp, to this electricity, you have to work so hard. You have to go to the middle of ocean and drill it and get out petroleum and... In this way your real business of life is finished."*
~ *Srila Prabhupada ( Conversation, West Virginia, June 24, 1976)*

have to reduce its profligate consumption of oil and/or cut back on other expenses, including weapons purchases.

The rising price of oil is producing what Pentagon contractor LMI calls a "fiscal disconnect" between the military's long-range objectives and the realities of the energy market place. "The need to recapitalize obsolete and damaged equipment [from the wars in Iraq and Afghanistan] and to develop high-technology systems to implement future operational concepts is growing," it explained in an April 2007 report. However, an inability "to control increased energy costs from fuel and supporting infrastructure diverts resources that would otherwise be available to procure new capabilities."

And this is likely to be the least of the Pentagon's worries. The Department of Defense is, after all, the world's richest military organization, and so can be expected to tap into hidden accounts of one sort or another in order to pay its oil bills and finance its many pet weapons projects. However, this assumes that sufficient petroleum will be available on world markets to meet the Pentagon's ever-growing needs — by no means a foregone conclusion. Like every other large consumer, the DoD must now confront the looming —

*These politicians, they create an atmosphere... Therefore I say the head of the state, they must be clean. But they are all motivated. Therefore the whole world is in chaotic condition. Generally politician has got a particular motive behind him. And when they cannot pull on they declare war. Pakistan, since the beginning, they could not make the economic condition very sound. But when the people were too much agitated, they declared war with India. The whole attention was... And they have been educated in such a way that India is their strongest enemy. Anything Indian, they dislike in Pakistan. So this is going on by the politicians. They are creating situation because they are not honest, they are not clean. ...and the fight is going on, and the poor people in the state, they are suffering.*
*-Srila Prabhupada (Room Conversation, with Richard Webster, May 24, 1974, Rome)*

but hard to assess — reality of "Peak Oil"; the very real possibility that global oil production is at or near its maximum sustainable ("peak") output and will soon commence an irreversible decline.

That global oil output will eventually reach a peak and then decline is no longer a matter of debate; all major energy organizations have now embraced this view. What remains open for argument is precisely when this moment will arrive. Some experts place it comfortably in the future — meaning two or three decades down the pike — while others put it in this very decade. If there is a consensus emerging, it is that peak-oil output will occur somewhere around 2015. Whatever the timing of this momentous event, it is apparent that the world faces a profound shift in the global availability of energy, as we move from a situation of relative abundance to one of relative scarcity. It should be noted, moreover, that this shift will apply, above all, to the form of energy most in demand by the Pentagon: the petroleum liquids used to power planes, ships, and armored vehicles.

## The Bush Doctrine Faces Peak Oil

Peak oil is not one of the global threats the Department of Defense has ever had to face before; and, like other U.S. government agencies, it tended to avoid the issue, viewing it until recently as a peripheral matter. As intimations of peak oil's imminent arrival increased, however, it has been forced to sit up and take notice. Spurred perhaps by rising fuel prices, or by the growing attention being devoted to "energy security" by academic strategists, the DoD has suddenly taken an interest in the problem. To guide its exploration of the issue, the Office of Force Transformation within the Office of the Under Secretary of Defense for Policy commissioned LMI to conduct a study on the implications of future energy scarcity for Pentagon strategic planning.

The resulting study, "Transforming the Way the DoD Looks at Energy," was a bombshell. Determining that the Pentagon's favored

*"To our grandfathers and grandchildren, the cavemen...."*
*(Rene Barjavel 1911 - 1985)*

97

strategy of global military engagement is incompatible with a world of declining oil output, LMI concluded that "current planning presents a situation in which the aggregate operational capability of the force may be unsustainable in the long term."

LMI arrived at this conclusion from a careful analysis of current U.S. military doctrine. At the heart of the national military strategy imposed by the Bush administration — the Bush Doctrine — are two core principles: transformation, or the conversion of America's stodgy, tank-heavy Cold War military apparatus into an agile, continent-hopping high-tech, futuristic war machine; and pre-emption, or the initiation of hostilities against "rogue states" like Iraq and Iran, thought to be pursuing weapons of mass destruction. What both principles entail is a substantial increase in the Pentagon's consumption of petroleum products — either because such plans rely, to an increased extent, on air and sea-power or because they imply an accelerated tempo of military operations.

As summarized by LMI, implementation of the Bush Doctrine requires that "our forces must expand geographically and be more mobile and expeditionary so that they can be engaged in more theaters and prepared for expedient deployment anywhere in the world"; at the same time, they "must transition from a reactive to a proactive force posture to deter enemy forces from organizing for and conducting potentially catastrophic attacks." It follows that, "to carry out these activities, the U.S. military will have to be even more energy intense…. Considering the trend in operational fuel consumption and future capability needs, this 'new' force employment

*"But what is the use of these big, big motorbuses and acquire petrol, machine, factory, so many things? But nature's way there is already means of transport. The horses are there. The bulls are there. But they will eat them, and they will create these motor big, big buses and then petrol, then fight…"*
*~Srila Prabhupada (Morning walk, Mauritius, 2 Oct, 1975)*

construct will likely demand more energy/fuel in the deployed setting."

The resulting increase in petroleum consumption is likely to prove dramatic. During Operation Desert Storm in 1991, the average American soldier consumed only four gallons of oil per day; as a result of George W. Bush's initiatives, a U.S. soldier in Iraq is now using four times as much. If this rate of increase continues unabated, the next major war could entail an expenditure of 64 gallons per soldier per day.

It was the unassailable logic of this situation that led LMI to conclude that there is a severe "operational disconnect" between the Bush administration's principles for future war-fighting and the global energy situation. The administration has, the company notes, "tethered operational capability to high-technology solutions that require continued growth in energy sources" — and done so at the worst possible moment historically. After all, the likelihood is that the global energy supply is about to begin diminishing rather than expanding. Clearly, writes LMI in its April 2007 report, "it may not be possible to execute operational concepts and capabilities to achieve our security strategy if the energy implications are not considered." And when those energy implications are considered, the strategy appears "unsustainable."

## Pentagon As A Global Oil-Protection Service

How will the military respond to this unexpected challenge? One approach, favored by some within the DoD, is to go "green" — that

*There will be no stamp commemorating year 1959-2059. Oil industry will not observe its bicentenary.*

is, to emphasize the accelerated development and acquisition of fuel-efficient weapons systems so that the Pentagon can retain its commitment to the Bush Doctrine, but consume less oil while doing so. This approach, if feasible, would have the obvious attraction of allowing the Pentagon to assume an environmentally-friendly facade while maintaining and developing its existing, interventionist force structure.

But there is also a more sinister approach that may be far more highly favored by senior officials: To ensure itself a "reliable" source of oil in perpetuity, the Pentagon will increase its efforts to maintain control over foreign sources of supply, notably oil fields and refineries in the Persian Gulf region, especially in Iraq, Kuwait, Qatar, Saudi Arabia, and the United Arab Emirates. This would help explain the recent talk of U.S. plans to retain "enduring" bases in Iraq, along with its already impressive and elaborate basing infrastructure in these other countries.

The U.S. military first began procuring petroleum products from Persian Gulf suppliers to sustain combat operations in the Middle East and Asia during World War II, and has been doing so ever since. It was, in part, to protect this vital source of petroleum for military purposes that, in 1945, President Roosevelt first proposed the deployment of an American military presence in the Persian Gulf region. Later, the protection of Persian Gulf oil became more important for the economic well-being of the United States, as articulated in President Jimmy Carter's "Carter Doctrine" speech of January 23, 1980 as well as in President George H. W. Bush's August 1990 decision to stop Saddam Hussein's invasion of Kuwait, which led to the first Gulf War — and, many would argue, the decision of the younger Bush to invade Iraq over a decade later.

> *"Petroleum's worth as a portable, dense energy source powering the vast majority of vehicles and as the base of many industrial chemicals makes it one of the world's most important commodities. Access to it was a major factor in several military conflicts in last one hundred years."*
> ~Joseph Moore

Along the way, the American military has been transformed into a "global oil-protection service" for the benefit of U.S. corporations and consumers, fighting overseas battles and establishing its bases to ensure that we get our daily fuel fix. It would be both sad and ironic, if the military now began fighting wars mainly so that it could be guaranteed the fuel to run its own planes, ships, and tanks — consuming hundreds of billions of dollars a year that could instead be spent on the development of petroleum alternatives.

# Recklessly Wasteful

### A Conversation With
### His Divine Grace A.C.Bhaktivedanta Swami Prabhupada

Prabhupada: I think that in Bible there is a story, prodigal son? So we are prodigal son. We are all sons of God, now we have become prodigal sons. What is the meaning of prodigal? "Without any responsibility," is it not? Do whatever you like.

Translator: Run away from the protection of the family.

Prabhupada: Yes. That is our position, that we are sons of God, we have given up protection of God. God is protecting in all circumstances.

Hari Sauri: It says "Prodigal: recklessly wasteful."

Prabhupāda: That's it. This is the... We are all recklessly wasteful sons of God. We are sons of God, there is no doubt, but at the present moment, recklessly wasteful. We are wasting our valuable life even, we are so reckless. So the Krishna consciousness movement is to check their recklessness and bring them into senses of responsibility, going back home, back to Godhead. This is Krishna consciousness. But people are so reckless, as soon as you say something of God, immediately they laugh, "Oh, what is nonsense, God." This is the supreme recklessness. India was very serious about God. Still, India is serious. Now, the present leaders, they are thinking that Indians are spoiled, simply thinking of God—they're not thinking like the Americans and Europeans for economic development. So this is the position, and it is very difficult, but still we can do something this to the humanity, by preaching this Krsna consciousness movement. And those who are fortunate, they'll come, take up seriously. These reckless prodigal sons, we have got so many examples. For example, just like there is some stock of petroleum and they got information that from petroleum they can run on cars without horse. So, manufacture millions of cars and spoil the whole oil. This is recklessness. And when it is finished, then they'll cry. And it will be finished. This is going on. Recklessness. Just as a reckless boy, father has left some property, use it, use it. As soon as you get. The sooner it is finished, that's all. That is recklessness. There is some strength in the body, and as soon as he gets a little taste of sex life, "Oh, spend it, spend it," whole energy spent. The brain becomes vacant. This is recklessness. Beginning from twelfth year, by the thirty year, everything finished. Then he's impotent. In our childhood—in our childhood means, say, eighty years ago, or say, a hundred years ago—there was no motorcar. And now, wherever you go, in any country, you see thousands and millions of car. This is recklessness. Hundreds years ago they could do without motorcar, and now they cannot live without a car. In this way, unnecessarily, they're increasing bodily or material necessities of life. This is recklessness.

*(Room Conversation With French Commander, August 3, 1976, France)*

*"People really feel that, when they go to the gas pump now, that the oil cartel is holding them by the legs and tipping them upside down and shaking money out of their pockets."*
-Ed Markey

# Section-IV

# Oil
## &
# Environment

*"Practically every environmental problem we have can be traced to our addiction to fossil fuels, primarily oil."*   *–Dennis Weaver*

There is a sanctity about earth. Even lifelong urban dwellers are revolted by lakes of oil, stacks of crunched automobiles, unclean air, stinking sewage systems, dying forests, ugly garbage dumps, and unswimmable lakes and rivers.

But what's the cause of this? A polluted environment grows out of polluted consciousness. It's like Gandhi said-there's enough on earth for everyone's need but not for everyone's greed.

We have witnessed the air pollution in Mexico, the deadly aftermath of the Chernobyl accident in the Soviet Union, the destruction of the rainforests in Brazil, Swedish lakes dead from acid rain, and the horror of the chemical disaster in Bhopal, India. Wherever we look, we see a planet in trouble, a planet in need of spiritual healing. It's not difficult to identify greed as a root cause of pollution as we briefly survey the state of the world's ecological predicament.

Mountains of rubbish have become symbolic of people who have more than they need. This is especially true in the industrialized nations, where the typical resident uses 10 times more steel, 12 times more fuel, and 15 times more paper than a typical resident of the developing nations.

## Oil's Environmental Impact

### Water Pollution

Comprising over 70% of the Earth's surface, water is undoubtedly the most precious natural resource that exists on our planet. Without the seemingly invaluable compound comprised of hydrogen and oxygen, life on Earth would be nonexistent: it is essential for everything on our planet to grow and prosper. Although we as humans recognize this fact, we disregard it by polluting our rivers, lakes, and oceans. Subsequently, we are slowly but surely harming our planet to the point where organisms are dying at a very alarming rate. In addition to innocent organisms dying off, our drinking water has become greatly affected as is our ability to use water for recreational purposes.

The world's oceans and rivers have never been under more pressure from pollution. One kind of water pollution, which is usually the most common, is called "Conventional" and is made up of conventional pollutants. Conventional pollutants are solid particles and matter found in our water. Most of the pollution we can see is

*Devotee: In Satya-yuga or Treta-yuga they would not drill for petrol to use to make electricity or to make these gas lamps, so by what means did they use to light the palaces and the kingdoms? What was the natural resource utilized for that purpose?*
*Prabhupada: This oil. You produce castor seed oil. You grow castor seed by agriculture. (Room Conversation, February 21, 1977, Mayapura)*

conventional. Cans, bottles, paper, just about anything, can be a conventional pollutant. Conventional pollutants cause a wide variety of environmental problems. The solids suspended in the water can block the sun's rays, and this blocking disrupts the carbon dioxide/oxygen conversion process. This process is vital to an aquatic food chain. Sometimes the solid pollution is so bad, the water becomes unusable to humans and animals. The best way to remove conventional pollutants is to run the water through a treatment plant. In treatment plants the water is skimmed, run through several filters, and settled. This removes about 60 percent of the pollutants. The remaining pollution is decomposed by tiny pollution-eating microorganisms.

Another type of pollution is called Non-conventional and is made up of non-conventional pollutants. Non-conventional pollutants are more dangerous to the environment than conventional pollutants. Non-conventional pollutants are dissolved metals, both toxic (harmful) and nontoxic (not harmful). Many factories dump these pollutants into the water as byproducts of their production process. The most devastating type of non-conventional pollution is an oil spill. More than 13,000 oil spills occur each year in the United States alone. Motor oil can damage or kill aquatic vegetation and animal life. An oil slick - contaminating two million gallons of drinking water - can develop from one quart of oil. Massive oil drilling mishaps are responsible for serious water pollution. For example, a 1979 accident at Ixtoc2, an oil well in the Gulf of Mexico, spilled 140 million gallons, covering 10 percent of the Gulf.

Transporting oil is also dangerous. In an average year, accidents dump about 120 million gallons of oil into the sea. But "roughly six times more oil gets into the ocean simply through routine flushing of carrier tanks, runoff from streets, and other everyday consequences of motor vehicle use," says Marcia D. Lowe of the World Watch Institute.

Another major pollution contributors are the meat-packing industry and many manufacturing industries. Estimates suggest that nearly 1.5 billion people lack safe drinking water and that at least 5 million deaths per year can be attributed to waterborne diseases. This phenomenon is not confined to the Third World. Even many of the municipal water supplies in developed countries can present health risks. The water in European nations such as Germany is also found to be contaminated with dangerous amounts of residues from the chemical, pharmaceutical, and metal-working industries. In 1992, Germany experienced 1,825 documented accidents involving releases of water-polluting substances.

In Japan, during 1953-60, there was a water borne epidemic called Minimata epidemic. It was a case of mercury poisoning caused by consumption of fish from the Minimata Bay of Japan which was heavily contaminated by mercury compounds discharged by a nearby plastic industry. It was characterized by severe damage to the nervous system leading to ataxia, paraesthesia (abnormal pricking sensations), loss of vision and hearing and ultimately death.

Another case of mercury poisoning was the epidemic of consumption of fish from polluted Kalu River in the Thana district of Bombay, India. The major symptom was paralysis. Such incidents are too numerous to mention.

### Air Pollution

Most of us do not realize that our air consumption is more than 10000 times than that of water in terms of volume. Motor vehicles and industries are the world's biggest source of air pollution.

The air we breathe in many cities is being polluted by driving cars and trucks; burning coal, oil, and other fossil fuels; and manufacturing chemicals. Millions of people live in areas where urban smog, very small particles, and toxic pollutants pose serious health concerns.

Carbon monoxide is an odorless, colorless gas and is formed when the carbon in fuels does not completely burn. Vehicle exhaust contributes roughly 60 percent of all carbon monoxide emissions nationwide, and up to 95 percent in cities. Other sources include fuel combustion in industrial processes and natural sources such as wildfires.

According to the Environmental Protection Agency, the average adult breathes over 3,000 gallons of air every day. Children breathe even more air per pound of body weight and are more susceptible to air pollution.

Air pollution threatens the health of human beings and other living things on our planet. While often invisible, pollutants in the air create smog and acid rain, cause cancer or other serious health effects, diminish the protective ozone layer in the upper atmosphere, and contribute to the potential for world climate change.

A study by the World Health Organization and the United Nations Environment Program shows that two thirds of the world's urban population live with polluted air.

Breathing the air for a day in Mumbai, India, pollutes your lungs as much as smoking ten cigarettes.

Air is the ocean we breathe. Air supplies us with oxygen which is essential for our bodies to live. Air is 99.9% nitrogen, oxygen, water vapor and inert gases. Human activities can release substances into the air, some of which can cause problems for humans, plants, and animals.

There are several main types of pollution and well-known effects of pollution which are commonly discussed. These include smog, acid rain, the greenhouse effect, and "holes" in the ozone layer. Each of these problems has serious implications for our health and well-being as well as for the whole environment.

One type of air pollution is the release of particles into the air from burning fuel for energy. Diesel smoke is a good example of this particulate matter . The particles are very small pieces of matter measuring about 2.5 microns or about .0001 inches. This type of pollution is sometimes referred to as "black carbon" pollution.

Another type of pollution is the release of noxious gases, such as sulfur dioxide, carbon monoxide, nitrogen oxides, and chemical vapors. These can take part in further chemical reactions once they are in the atmosphere, forming smog and acid rain.

## Oil Spills

An oil spill is the release of a liquid petroleum hydrocarbon into the environment as a result of human activity (can be both intentional and unintentional). The term often refers to marine oil spills, where oil is released into the ocean or coastal waters. Oil can refer to many

different materials, including crude oil, refined petroleum products (such as gasoline or diesel fuel) or by-products, ships' bunkers, oily refuse or oil mixed in waste. Spills take months or even years to clean up. Of course, chemicals used in cleaning further poison the environment and all of the oil can never really be cleared.

Between 1978 and 1991, prior to the Persian Gulf War, five major oil spills had occurred in the Gulf, each involving more than a quarter of a million barrels of crude oil and each being larger than the 1989 Exxon Valdez spill. The largest of these spills was associated with a well at Nowruz, Iran that resulted in 1.9 million barrels of oil being dumped in the northern section of the Gulf. Also, a considerable amount of industrial spillage and natural oil seepage occurs in the Gulf each year. Estimates range from 250,000 to 3 million barrels per year (DeSouza, 1991; Ackleson et al., 1992). This is the environmental price, which the Gulf must pay to be the world's major oil highway.

The Exxon Valdez Oil Spill, which occurred on 24 March 1989, is considered one of the most devastating man-made environmental disasters ever

*"Socialism collapsed because it did not allow the market to tell the economic truth. Capitalism may collapse because it does not allow the market to tell the ecological truth."*
*~Oystein Dahle, former vice president of Exxon*

to occur at sea. As significant as the Exxon Valdez spill was, it ranks well down on the list of the world's largest oil spills in terms of volume released. The region was a habitat for salmon, sea otters, seals, sea birds and the great white shark. The vessel spilled 10.8 million gallons of unrefined Alaskan crude oil into the sea.

In an average year, accidents dump about 120 million gallons of oil into the sea. But "roughly six times more oil gets into the ocean simply through routine flushing of carrier tanks, runoff from streets, and other everyday consequences of motor vehicle use," says Marcia D. Lowe of the World Watch Institute.

A fatal system error has occurred in the Gulf. Continue?
OK    Cancel

## Acid Rain

Acid rain is a result of air pollution. When any type of fuel is burnt, lots of different chemicals are produced. The smoke that comes from a fire or the fumes that come out of a car exhaust don't just contain the sooty grey particles that you can see - they also contains lots of invisible gases that can be even more harmful to our environment.

Power stations, factories and cars all burn fuels and therefore they all produce polluting gases. Some of these gases (especially nitrogen oxides and sulphur dioxide) react with the tiny droplets of water in clouds to form

Acid Rain Formation

$SO_2$    $NO_x$    Acid Rain

sulphuric and nitric acids. The rain from these clouds then falls as very weak acid - which is why it is known as "acid rain".

Acidity is measured using a scale called the pH scale. This scale goes from 0 to 14. 0 is the most acidic and 14 is the most alkaline (opposite of acidic). Something with a pH value of 7, we call neutral, this means that it is neither acidic nor alkaline.

Very strong acids will burn if they touch your skin and can even destroy metals. Acid rain is much weaker than this, never acidic enough to burn your skin.

Rain is always slightly acidic because it mixes with naturally occurring oxides in the air. Unpolluted rain would have a pH value of between 5 and 6. When the air becomes more polluted with nitrogen oxides and sulphur dioxide the acidity can increase to a pH value of 4. Some rain has even been recorded as being pH2.

Vinegar has a pH value of 2.2 and lemon juice has a value of pH2.3. Sometimes acid rain is as acidic as lemon juice or vinegar. Acid rain can be carried great distances in the atmosphere, not just between countries but

*Prabhupda: So you ask problem, I will answer. Your energy, problems of energy, petrol, it will be automatically solved. If we are localized, there is no question of petrol.*

*Bhagavan: You say in the, I think it's in the Second Canto of Srimad-Bhagavatam, that by doing so much drilling into the earth, they actually disturbed the rotation of the earth.*

*Prabhupada: Yes, we can think like that. Just like the plane, aeroplane, is flying. There is sufficient petrol stock. Is it not? So the world has got sufficient petrol stock. If you do not know how it is being used, maybe due to this petrol, it is floating. And if you take away the petrol stock, it may drop. Everything is there. There is a purpose. Purnam idam. [Isopanisad, Invocation]: "The Personality of Godhead is perfect and complete, and because He is completely perfect, all emanations from Him, such as this phenomenal world, are perfectly equipped as complete wholes..."] There is full purpose. Not that whimsically petrol is there within the earth. There is some purpose.*

*Devotee: What they do, Srila Prabhupāda, is take the petrol out and put salt water, because they know there can be a imbalance. And then they put salt water in the holes.*

*Prabhupada: But water cannot produce gas. Petrol produces gas. Maybe due to that gas, it is floating. Because we have got practical experience. When there is gas, you can float anything. (Morning walk, May 27, 1974, Rome)*

also from continent to continent. The acid can also take the form of snow, mists and dry dusts. The rain sometimes falls many miles from the source of pollution but wherever it falls it can have a serious effect on soil, trees, buildings and water.

Forests all over the world are dying, fish are dying. In Scandinavia there are dead lakes, which are crystal clear and contain no living creatures or plant life. Many of Britain's freshwater fish are threatened, there have been reports of deformed fish being hatched. This leads to fish-eating birds and animals being affected also.

Since the Industrial Revolution, emissions of sulfur dioxide and nitrogen oxides to the atmosphere have increased. Acid rain was first found in Manchester, England. In 1852, Robert Angus Smith found the relationship between acid rain and atmospheric pollution. Though acid rain was discovered in 1852, it wasn't until the late 1960s that scientists began widely observing and studying the phenomenon.

In western Europe, millions of hectares of forests have minor or major damage from acid rain.

Lakes and rivers can have powdered limestone added to them to neutralize the water - this is called "liming". Liming, however, is expensive and its effects are only temporary - it needs to be continued until the acid rain stops. The people of Norway and Sweden have successfully used liming to help restore lakes and streams in their countries.

## Global Warming

Global warming is the increase in the average temperature of the Earth's near-surface air and oceans in recent decades and its projected continuation.

Carbon dioxide makes up only .03 percent of the atmosphere, but it is extremely important. Scientists say it traps heat that otherwise would escape into space. This greenhouse effect keeps the earth's climate from becoming uncomfortably cold.

Carbon dioxide is a natural product of organic decay and animal respiration. But industry has poured additional carbon dioxide into the air, causing potentially dangerous increases in the earth's temperature.

The global average air temperature near the Earth's surface rose 0.74 ± 0.18 °C (1.33 ± 0.32 °F) during the 100 years ending in 2005. Climate model projections summarized by the IPCC indicate that average global surface temperature will likely rise a further 1.1 to 6.4 °C (2.0 to 11.5 °F) during the 21st century.

About 75 percent of the carbon dioxide entering the atmosphere each year comes from the burning of fossil fuels in factories and motor vehicles.

Another 20 percent comes from the deliberate burning of forests to clear land.

United Nations studies show that a warming climate could raise sea levels, which are already rising, by 1.5 to 6.5 feet over the next century. If sea levels rise 1 meter, this could submerge 5 million square kilometers of lowlands. These lowlands are now inhabited by 1 billion people and include one third of the world's cropland.

Not everyone may agree with such doomsday scenarios but among the scientists who have deliberated on the issue for years and have drawn different conclusions from their respective studies, have one thing in common that global warming is a fact and that rising sea levels remain a potential source of worry.

## Ozone Layer Destruction

Ozone is a form of oxygen that forms high up in the atmosphere when sunlight breaks apart oxygen atoms. Most ozone in the atmosphere is found between 19 and 30 km above the Earth's surface where it forms the ozone layer. The ozone layer acts as the planet's sunscreen, filtering out harmful UV rays coming from the Sun, which have many impacts at the surface, including damage to our health, wildlife and certain materials.

Since the 1970s we have realised that CFCs, man-made chemicals used in refrigeration have been depleting ozone and damaging the ozone layer. The damage is worst over Antarctica, where a large ozone hole forms every spring in September and October.

The ozone layer filters out incoming radiation in the "cell-damaging" ultraviolet (UV) part of the spectrum. Without ozone, life on Earth would not have been the way it is. The discovery of a large ozone hole over Antarctica and its association with man-made CFCs led the world to take action to protect the ozone layer.

CFC is short for chlorofluorcarbon, a man-made chemical made up of carbon, chlorine and fluorine atoms. There are many different types of CFCs. They were first developed in the 1930s, and since then have been widely used in refrigerators, aerosol cans and fire extinguishers. Man-made CFCs have been the main cause of ozone depletion high up in the atmosphere. One CFC molecule can destroy up to 100,000 ozone molecules. To date, they have accounted for roughly 80% of the total ozone depletion that has been observed. Other man-made chemicals like halons have accounted for the rest. Fortunately, the use of new CFCs has been banned since 1995.

UV rays are damaging to our health. Prolonged exposure to the Sun causes sunburn, which over many years can develop into skin cancer. UV rays can also damage eyesight. UV rays are also harmful to other forms of wildlife, particularly small plants and animals living in the sea called plankton. Plankton are not protected from the Sun's rays, which can penetrate water to a depth of many meters. Plankton form the base of the ocean food chain. UV rays can damage certain crops, like rice, which many people in the world rely on for food. Finally, UV rays can damage paint, clothing and other materials.

Any decrease in the amount of ozone in the ozone layer will increase the amount of UV rays reaching the Earth's surface, and worsen the impacts due to UV exposure.

### Environmental Warfare

Environmental warfare is defined as the intentional modification of a system of the natural ecology, such as climate and weather systems to cause intentional economic, psycho-social or physical destruction of an intended target, geophysical or population location, as part of strategic or tactical war.

Environmental warfare consists of the deliberate and illegal destruction, exploitation, or modification of the environment as a strategy of war or in times of armed conflict (including civil conflict within states). Modification of the environment that occurs during armed conflict and is likely to have widespread, long-lasting, or severe effects is proscribed by the Convention on the Prohibition of Military or Any Other Hostile Use of Environmental Modification Techniques, adopted by the United Nations General Assembly in 1976. Nevertheless, such destruction has occurred with some regularity. In the 1960s and '70s, the U.S. military used the defoliant Agent Orange to destroy forest cover in Vietnam, and in 1991 Iraqi military forces retreating during the Persian Gulf War set fire to Kuwaiti oil wells, causing significant environmental damage. The Rome Statute of the International Criminal Court, adopted in 1998, defines such modification or destruction as a war crime.

In the Gulf War of 1991, Iraq, it appears, resorted to environmental warfare, deliberately releasing millions of gallons of oil from Kuwaiti fields into the Persian Gulf, perhaps as a defense against amphibious assault, or perhaps as a means of crippling Saudi Arabia's water desalinization plants. Also, Iraq apparently set hundreds of oil wells ablaze, releasing clouds of black smoke that turned day into night over much of Kuwait.

Environmental warfare may sound new to some, but it has been researched extensively in military circles for years. The first public description of weather modification techniques as a weapon of war was made on 20 March, 1974. At that time the Pentagon revealed a seven-year cloud seeding effort in Vietnam and Cambodia, costing $21.6 million. The objective was to increase rainfall in target areas, thereby causing landslides and making unpaved roads muddy, hindering the movement of supplies. (*Science, "Weather Warfare: Pentagon Cencedes 7-Year Vietnam Effort," Deborah Shapley, June 7, 1974.*) But interest in the exploitation of the environment for military purposes did not end there.

Air University, located at Maxwell Air Force Base in Alabama, describes itself as a "center for advanced education" that "plays a vital role in fulfilling the mission of the United States Air Force" and whose "service members must place the nation's defense above self." The Chief of Staff of the US Air Force tasked Air University to "look 30 years into the future to identify the concepts, capabilities and technologies the United States will require to remain the dominant air and space force in the 21st century." The study, completed in 1996, was titled 'Air Force 2025'. One component of the study was a paper titled 'Weather as a Force Multiplier: Owning the Weather in 2025.' It is a chilling document. It is evident that the authors regard our environment as nothing more than a resource to be exploited for military purposes. They claim that by 2025 US forces can "own the weather" by "capitalizing on emerging technologies and focusing development of those technologies to warfighting applications." The authors describe weather modification as having "tremendous military capabilities" which "can provide battlespace dominance to a degree never before imagined," claiming the project would be "not unlike the splitting of the atom." The paper goes on to discuss how ionospheric research (The ionosphere is a region of the earth's atmosphere ranging from about 30 – 1200 miles above the surface of the earth.) is necessary to achieve goals in both enhancing US communication capabilities and as a method of disabling enemy communications. "By 2025, it may be possible to modify the ionosphere and near space, creating a variety of potential applications."

Dr. Bernard Eastlund, while working as a consultant for Advanced Power Technologies Inc. (APTI) in the 1980s, patented devices that are described as capable of "causing…total disruption of communications over a very large portion of the Earth…missile or aircraft destruction, deflection or confusion…weather modification…" (*From the book 'Angels Don't Play This HAARP')* These patents were based on the ideas and fundamental research of Nicola Tesla (many of his ideas were stolen by US corporations). Some of Eastlund's patents were temporarily sealed under a US Secrecy Order. APTI and Eastlund's patents were quickly purchased by E-Systems, a company that is home to many retired and currently employed CIA agents. In 1993 E-Systems received $1.8 billion in classified contracts. Raytheon, the fourth largest US defense contractor and third largest aerospace company, currently holds the patents.

In light of the above, it is significant to know that since the early 1990s, the US Air Force has been sponsoring the world's largest ionospheric modification project called HAARP (High-frequency Active Auroral Research Program). HAARP, located in the remote bush country of Gakona, Alaska. The military was interested because, in the event of a Russian nuclear attack on the United States, an Alaskan site would be under the path of the incoming warheads." HAARP is currently a part of the ongoing Strategic Defense Initiative (SDI), popularly known as "Star Wars". ("*Natural Disasters and Meteorological War?" by Vera Vratusa*)

## Indirect Impact

Oil is the foundation upon which the industrial civilization is built and around which it operates. So oil has to share some responsibility for all the environmental catastrophes occurring today.

## Wild Life Damage

Studies of the Exxon Valdez oil spill have shown that the environmental damage caused by oil spills can be greater than was previously thought. Petroleum-based hydrocarbons can negatively impact marine life at concentrations as low as one part per billion.

The lighter fractions of oil, such as benzene and toluene, are highly toxic, but are also volatile and evaporate quickly. Heavier components

*Wild animals are just as confused as people are now. You've got toxins in the water, oil, sewage, all sorts of things.*
*~Jack Hannah*

of crude oil, such as polycyclic aromatic hydrocarbons (PAHs) appear to cause the most damage; while they are less toxic than the lighter volatiles, they persist in the environment much longer. A heavy oil spill can also blanket estuaries and shoreline ecosystems such as salt marshes and tidal pools, preventing gas exchange and blocking light. The oil can mix deeply into pebble, shingle or sandy beaches, where it may remain for months or even years.

Seabirds are severely affected by spills as the oil penetrates and opens up the structure of their plumage, reducing the insulating ability of their feathers, making the birds more vulnerable to temperature fluctuations and much less buoyant in the water. The oiled feathers also impairs birds' flight abilities, making it difficult or impossible to forage and escape from predators. As they attempt to preen, birds typically ingest oil that coats their feathers, causing kidney damage, altered liver function, and digestive tract irritation. The limited foraging ability coupled with the ingestion of the oil quickly causes dehydration and metabolic imbalances. Most birds affected by an oil spill will die without human intervention.

Marine mammals exposed to oil spills are affected in many of the same ways as seabirds. Oil coats the fur of Sea otters, seals, reducing their furs natural insulation abilities, leading to body temperature fluctuations and hypothermia. Ingestion of the oil also causes dehydration, and impaired digestion.

Once part of the largest intact wilderness area in the United States, Alaska's North Slope now hosts one of the world's largest industrial complexes, spanning some 1,000 square miles of once-pristine Arctic

*Prabhupada: Petrol smelling everywhere.*
*Pusta Krishna: It's from these automobiles, the exhaust. They say that in some cities like New York, just living in the city itself, it is like smoking two packs of cigarettes every day because of so much pollution in the air, so contaminated.*
*(Morning Walk, June 8, 1976, Los Angeles)*

tundra. All of this industrial activity is taking place in an exceptionally fragile region. Because of the very short summer growing season, extreme cold at other times of the year, and nutrient-poor soils and permafrost, vegetation grows very slowly in the North Slope. Any physical disturbance -- bulldozer tracks, seismic oil exploration, spills of oil and other toxic substances -- can scar the land for decades.

Oil related pollution is damaging life forms, both on water and land. On various highways around the world, many creatures get killed or maimed by speeding vehicles. Just in one state of America, Pennsylvania, deer-vehicle accidents account for more than $130 million in insurance claims. In addition, it is estimated that more than 150,000 deer-vehicle collisions likely occur in New York and Pennsylvania annually.

In India, in the last decade, several dozens of elephants were killed and many more injured by speeding trucks and trains, mainly in foothills areas.

## Trash

Disposing cars is another problem. Heaps of smashed cars is a common sight and an increasing problem all over the world.

The annual disposal of about 8 - 9 million cars in Europe has severe negative impacts on the environment. In addition to the environmental problems of car use, the disposal of cars is a major source of hazardous waste and toxic emissions.

Current waste management of end-of-life vehicles is focused on the recovery of ferrous metals in steel works. Cars are shredded and then separated into various fractions but only the metal fractions can be recycled. The remaining non-recyclable light fraction amounts to approximately 2 million tons of

hazardous waste annually, which is equivalent to 10% of the total amount of hazardous wastes generated in the European Union. This waste is contaminated with heavy metals, polyvinyl chloride (PVC) plastic, plasticizers and hazardous oils. Most of this is disposed of in landfills, where leaching can lead to contamination of soil and groundwater. The presence of PVC plastics and other chlorinated materials poses particular hazards, such as dioxin formation, if this waste is incinerated, or accidentally catch fire. PVC plastic residue in steel scrap also leads to dioxin emissions from steel recycling plants. Almost none of the PVC parts used in cars can be recycled.

The huge amounts of hazardous wastes and the toxic emissions generated by the disposal of end-of-life vehicles illustrate the environmental problems that arise in an industrialized setup.

Our trash is not only taking up land but is destroying resources, animals and their way of life.

## Toxic Waste

Lot of chemicals and processes which leave hazardous waste are intimately related to petroleum. Toxic waste is a waste material, often in chemical form, that can cause death or injury to living creatures. It usually is the product of industry or commerce, but comes also from residential use, agriculture, the military, medical facilities, radioactive sources, and light industry, such as dry- cleaning establishments. As with many pollution problems, toxic waste began to be a significant issue during the industrial revolution. The term is often used interchangeably with "hazardous waste," or discarded material that can pose a long-term risk to health or environment. Toxins can be released into air, water, or land.

Toxic waste can pollute the natural environment and contaminate groundwater. In US, Love Canal is a famous incident in which homes and schools were built near an area where toxic waste had been dumped, causing epidemic health problems.

A number of toxic substances that humans encounter regularly may pose serious health risks. Pesticide residues on vegetable crops, mercury in fish, and many industrially produced chemicals may cause cancer, birth defects, genetic mutations, or death.

Exotic chemicals such as dioxins and PCBs are not the only source of danger. Common heavy metals like lead, nickel, mercury, chromium, and cadmium all have poisonous effects on humans. For example, lead, found in old house paint and water pipes, is known to cause anemia, decreased intelligence, and other health problems in children. When trash is burned, poisonous heavy metals go into the air, and when trash is buried in landfills the heavy metals often migrate into human drinking water.

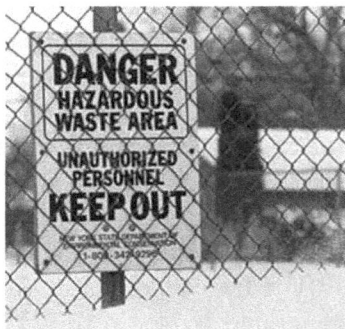

Bubbling, oozing, slimy, stinky, the United States has at least 36,000 hazardous waste sites containing used oil, battery acid, PCB's, heavy metals, detergents, pesticides, old paint, plastics, radioactive wastes and more. In 1980, the United States Environmental Protection Agency (EPA) set up a "Superfund" for cleaning up the country's most hazardous toxic waste sites. Superfund sites are the nation's worst toxic waste sites: 1,305 are scheduled for cleanup on the National Priorities List (NPL). So far only 35 appear to have been cleared. Sometimes toxins stored at some of these sites leak out of their containers and make their way into ground water, soil, lakes and rivers. Accidents, such as train derailments or crashes involving tanker trucks transporting toxic wastes or oil spills from ships, can also release large quantities of these dangerous chemicals into the environment. Build-up of toxins can also result from release of small amounts over a large area and/or a long time. An example of this is the runoff of excessive fertilizers and pesticides applied to lawns and farm fields. Cleaning up these toxic wastes has cost $20 billion over the past 15 years in the United States. Cleanup of the rest of the toxins already in the environment (in the U.S.A.) is expected to cost between $100 and $500 billion (that is about $500 to $2500 per person in the United States.)

About 11 million people in the U.S., including 3-4 million children, live within 1 mile of a federal Superfund site and face potential public health risks. As of 1993, Germany had 140,000 known sites where toxic wastes had been dumped illegally, and officials estimated another 240,000 sites unknown. Three Superfund toxic waste sites in and around New Orleans were flooded by Hurricane Katrina and one remains underwater, Environmental Protection Agency officials suspect that hazardous materials are leaching into the environment.

In November 1998, nearly 3,000 tons of Taiwanese toxic waste were dumped in a field in the southern port of Sihanoukville, Cambodia. Dumped in an open field, the waste was scavenged by poor villagers, many of whom later complained of sickness; several died. Local people panicked: thousands fled the city.

Another problem with the chemical industry is the occurrence of major accidents. In 1985, a valve broke at the Union Carbide chemical plant at Bhopal, India, allowing 30 tons of lethal methyl isocyanate gas to escape. More than 3,000 people living nearby were killed, and another 17,000 received permanent injuries. Many others died later on.

Gujarat, India, is a state hugely affected with fluoride contamination that poses major problem for drinking water. With the water intensive industries — like chemicals, dye, pharmaceutical and textile—- concentration in the State, an irreparable damage to the environment has been caused by disposing toxic waste.

We can conclude our discussion here with one last fact: over 1.1 billion pounds of pesticides costing over 25 billion dollars are used every year in the United States, the world's leading consumer of pesticides. That's over four pounds of pesticide per person!

Such dangers should make us think whether do we really need industries to lead a sane and happy life.

## Toxic Trade

Greenpeace has documented hundreds of cases where developed countries have traded or transferred toxic waste problems to developing countries. Instead of receiving clean technologies, too often developing countries receive toxic waste, products and technologies.

This type of trade is immoral and environmentally destructive for the receiving countries and their people. It also prevents developed countries from investing in real solutions to pollution, and developing future markets in more appropriate technologies or products.

The most blatant offence has been the export of toxic wastes from developed to developing countries. Greenpeace has sought a ban on this type of toxic trade and achieved it through an international treaty called the Basel Convention. The convention came into force in 1992 but it was a weak treaty.

In August 1986, the city of Philadelphia (USA) loaded 15,000 tons of toxic ash from its trash incineration plant onto an oceangoing freighter, the Khian Sea. This ship spent 18 months in the Caribbean, looking unsuccessfully for a place to dump its dangerous cargo. Five continents and three name changes later, the ship allegedly found a place to legitimately offload the toxic ash, but Greenpeace says the waste was actually dumped illegally in the Indian Ocean in November of 1988.

In January 2008, in the wake of the Christmas electronic gadget buying season, with many buying new flat screen TVs, cell phones and computers and facing the problem of disposing the old ones, the Basel Action Network (BAN) and the Electronics TakeBack Coalition (ETBC) cautioned consumers not to be fooled by the majority of businesses nationwide calling themselves electronics recyclers who in fact don't do any recycling at all, but instead ship old equipment to developing countries.

India gets ships from countries like Bermuda, Panama and even land-locked Mongolia to dismantle at Asia's biggest ship breaking yard, Alang. But scratch the surface and it appears that these small countries are a front for rich nations to send their dirty cargo ships to India. The poor countries come handy for countries like Germany and Greece to circumvent international laws. The catch being that international laws prevent 'the rich' from shipping their hazardous waste directly to India. 'The poor' are not barred from dealing in this lucrative international scrap trade with another 'poor country'. The "Blue Lady" was more than a 2,000 plus capacity luxury cruise liner run by the world famous Star Cruise Limited. But after it became unusable, it shipped through various countries before being 'officially' sold to a company based in Liberia for a mere US $10. Sailing under the Liberian flag, it was sold to an Indian company.

## Pesticides

The idea of striving to create life-giving foods while simultaneously dousing them with deadly poisons is inelegant and illogical.

To keep profit margins high, factory farming routinely uses chemical pesticides to protect crops from insects and animals. A pesticide may be a chemical substance, biological agent (such as a virus or bacteria), antimicrobial, disinfectant or device used against any pest. Pests include insects, plant pathogens, weeds, mollusks, birds, mammals, fish, roundworms and microbes.

Over 98% of sprayed insecticides and 95% of herbicides reach a destination other than their target species, including nontarget species, air, water, bottom sediments, and food. Pesticide contaminates land and water when it escapes from production sites and storage tanks, when it runs off from fields, when it is discarded, when it is sprayed aerially, and when it is sprayed into water to kill algae.

The many good biological, nonchemical, and nontoxic methods are seldom used. The pesticide industry vigorously markets its products and promises vast savings for growers.

Insecticides are not crop-pest specific. They pose a serious risk to farm labour and farm animals, and most other animals in the environment as well. They all suffer from insecticide poisoning, by dermal contact, inhalation or ingestion. The lack of target specificity and an indiscriminate and excessive insecticide application has disturbed large components of biodiversity of agricultural lands.

Most insecticides are not completely degraded and so leave residues in the food, feed, other agricultural produce, soil and water. These residues are poisonous when the food or feed from sprayed crops are consumed without proper cleaning. In the process entire natural food chain gets contaminated.

The nature and intensity of insecticidal toxicity depends upon the chemical structure of the insecticide and the mode of its action and not on whether it is natural or synthetic. The level of exposure and the concentration of the insecticide or its residue in the body are critical factors.

In developed countries, many dangerous pesticides are banned but their manufacturing for export purposes is allowed. Almost one third of the pesticides exported by US companies are banned for use within US. But what goes out, comes back in. Many of the fruits and vegetables imported in US are sprayed with the very same banned pesticides.

Most of the pesticides contribute to global warming and the depletion of the ozone layer.

In the United States, pesticides were found to pollute every stream and over 90% of wells sampled in a study by the US Geological Survey. Pesticide residues have also been found in rain and groundwater. Studies by the UK government showed that pesticide concentrations exceeded those allowable for drinking water in some samples of river water and groundwater.

The World Health Organization and the UN Environment Programme estimate that each year, 3 million workers in agriculture in the developing world experience severe poisoning from pesticides, about 18,000 of whom die. According to one study, as many as 25 million workers in developing countries may suffer mild pesticide poisoning yearly. These are associated with acute health problems for workers that handle the chemicals, such as abdominal pain, dizziness, headaches, nausea, vomiting, as well as skin and eye problems. Additionally, many studies have indicated that pesticide exposure is associated with long-term health problems such as respiratory problems, memory disorders, dermatologic conditions, cancer, depression, neurologic deficits, miscarriages, and birth defects.

Children have been found to be especially susceptible to the harmful effects of pesticides. A number of research studies have found higher instances of brain cancer, leukemia and birth defects in children with early exposure to pesticides, according to the Natural Resources Defense Council.

## Soil Depletion And Damage

Pesticides and chemical fertilizers are wreaking havoc on topsoil in the fields around the world. Franklin Roosevelt said, "A nation that destroys its soil destroys itself." And like the old bumper sticker says, "Stop treating your soil like dirt."

In recent years the problem of soil contamination has become increasingly acute. Soil contamination is frequently caused by spillage of petroleum hydrocarbons from oil wells or storage tanks, or leakages from cross-country piping and storage networks utilized in the oil and process industries. Inland oil spills and leakages can cause localized death of agricultural and natural vegetation and extensive soil damage.

Accidental influxes of salts to the soil can devastate most vegetation in the spillage area and stop biological activity in a matter of days, rendering the land unproductive for farming or growth of any type of vegetation. Whether the high influxes of sodium chlorides in the soil result from years of irrigation, brackish water usage or an industrial process-related accident, the results are the same: almost all vegetation within the salt-contaminated area dies, and this in turn may also adversely affect the surrounding ecological balance.

Atina Diffley says, "All life depends on soil. When you kill the soil by destroying the biological life and organic matter, it is no longer soil, it is dirt."

## Some 'Oil Vs. Environment' Case Studies

### Ecuador

Fifty percent of Ecuador's national budget is funded by oil earnings and continued oil exploration and production is thought to be necessary to ensure the countries' well being. The country plans to increase production and holds auctions to increase foreign investment. Dependence on oil revenue has hindered Ecuador's environmental enforcement, which in turn has caused damaging consequences to indigenous tribes living in the Amazon region and to the environment in the eastern part of the country. The Indians of Ecuador, located in the Amazon region of Oriente, have joined forces for the past 20 years to resist oil exploration and demand rights to their ancestral lands.

Many of the indigenous tribes in the Amazon region that once numbered in the thousands have been reduced to the hundreds as a result of the pollution generated by oil exploration and other assaults.

Water contamination has led to increased risks of cancer, abortion, dermatitis, fungal infection, headaches, and nausea. Their drinking, bathing, and fishing water contain toxins much higher than the safety limits set by the US Environmental Protection Agency.

The oil companies that drilled in the rain forest were responsible for "felling thousands of acres of trees, dynamiting the earth, spilling vast amounts of oil, destroying habitats, and fouling rivers." Fish have died from water pollution and the game the tribes once hunted have retreated deeper into the jungle as a result of the deforestation. The Rainforest Action Network found that Texaco alone spilled 17 million gallons of crude oil, abandoned hundreds of unlined toxic waste ponds, and constructed oil roads that opened more than 2.5 million acres of the forest to colonization. As a result, Ecuador's rain forests are being cut down by oil companies and settlers at a rate of approximately 340,000 hectares a year. The wood is used for construction, roads, fuel and furniture.

The exploration for oil has created numerous environmental problems of all types in the Amazon region. The Amazon basin in Ecuador has the greatest number of plant species of any South American country. The Sierra highlands have been almost completely deforested. Also, the Oriente is a species rich jungle with numerous mammals in danger of extinction. Oil that was placed on roads to cut dust has flowed into rivers. Oil waste in the past was placed in holes in the ground that contaminated the forests and the rivers. Ecuadorian officials estimate that ruptures to the major pipeline alone have discharged more than 16.8 million gallons of oil into the Amazon over the past eighteen years (compared to the 10.8 million-gallon Exxon Valdez spill).

### Nigeria

Oil has been an important part of the Nigerian economy since vast reserves of petroleum were discovered in Nigeria in the 1950s. Revenues from oil were 219 million Naira in 1970 and it increased to 10.6 billion Naira in 1979. Shell Oil operates many of its oil facilities in the oil-rich Delta region of Nigeria. The Ogonis, an ethnic group that predominates in the Delta region, has protested that Shell's oil production has not only devastated the local environment, but has destroyed the economic viability of the region for local farmers and producers. The Nigerian Federal Government, on the other hand, has been charged with failing to enact and enforce environmental protections against oil damage by Shell and other oil companies. Furthermore, many Ogonis have been harassed and even killed by the Federal government for organizing protests and threatening sabotage of oil facilities.

Oil production in Nigeria has had severe environmental and human consequences for the indigenous peoples who inhabit the areas surrounding oil extraction. Nigeria's export of 12 million barrels of oil a day comes from 12% of the country's land, and indigenous minority communities in these areas receive no economic benefits. Indigenous groups are actually further impoverished due to environmental degradation from oil production and the lack of adequate regulations on multinational companies, as they become more vulnerable to food shortages, health hazards, loss of land, pollution, forced migration and unemployment.

The social and environmental costs of oil production have been extensive. They include destruction of wildlife and biodiversity, loss of fertile soil, pollution of air and drinking water, degradation of farmland and damage to aquatic ecosystems, all of which have caused  serious health problems for the inhabitants of areas surrounding oil production. Pollution is caused by gas flaring, above ground pipeline leakage, oil waste dumping and oil spills. Approximately 75% of gas produced is flared annually causing considerable ecological and physical damage to other resources such as land/soil, water and vegetation. Gas flares, which are often situated close to villages, produce soot which is deposited on building roofs of neighboring villages. Whenever it rains, the soot is washed off and the black ink-like water running from the roofs is believed to contain chemicals which adversely effect the fertility of the soil. Gas pipelines have also caused irreparable damage to lands once used for agricultural purposes. These pipes should be buried to reduce risk of fracture and spillage.

However, they are often laid above ground and run directly through villages, where oil leaks have rendered the land economically useless.

Oil spills and the dumping oil into waterways has been extensive, often poisoning drinking water and destroying vegetation. According to an independent record of Shell's spills from 1982 to 1992, 1,626,000 gallons were spilt from the company's Nigerian operations in 27 separate incidences. Of the number of spills recorded from Shell - a company which operates in more than 100 countries - 40% were in Nigeria.

Shell is also being accused of engaging in "widespread ecological disturbances, including explosions from seismic surveys, pollution from pipe-line leaks, blowouts, drilling fluids and refinery effluents, and land alienation and disruption of the natural terrain from construction of industry infrastructure and installations". For example, oil spill contamination of the top soil has rendered the soil in the surrounding areas "unsuitable for plant growth by reducing the availability of nutrients or by increasing toxic contents in the soil." Gas flaring, on the other hand, "has been associated with reduced crop yield and plant growth on nearby farms, and disruption of wildlife in the immediate vicinity". Shell and other oil companies have developed an easy and inexpensive way to deal with by-products from oil drilling: "indiscriminate dumping".

## Colombia

The Andes mountains in Colombia have become the newest oil hot spot with several international companies drilling in the region. Oil drilling is a profitable business for the exploring corporation as well as the Colombian government which receives a large sum of money for each barrel of oil recovered. However, the process is not without violence nor criticism from environmental groups. The

> *Krishna says, you wanted material world to enjoy. "All right, do it, here is material world. Take as much petrol as you like and drive motorcar and create accident, do, go on. But now I am giving you good advice, that give up this business, come back to Me."*
> *- Srila Prabhupada (Evening Darsana, August 11, 1976, Tehran)*

Marxist guerrillas repeatedly interrupt production through the use of terrorist tactics including bombings and kidnappings. Environmental groups challenge Colombian laws regarding environmental degradation due to the methods of oil exploration and extraction primarily caused by foreign corporations.

In 1993, British Petroleum and its partners located oil beneath the eastern plains in the Andes Mountains. The company predicted "it could be worth $3 billion a year in exports - the government hopes for $5 billion - by 1997." The Colombian government also commissioned British Petroleum to continue exploring for oil in its frontier areas in 1995.

Not everyone in Colombia is pleased that the oil companies are drilling more and more in their country. The drug cartels, peasant groups and paramilitary groups have wrecked havoc on the oil-pumping stations. Many of the oil wells are located in the "stomping ground" of the Medellin drug cartel as well as its competitors. The area is also the home to "the less publicized `emerald wars' (Colombia produces 60 percent of the world emerald supply), to three separate groups of Marxist guerrillas and to an increasingly terroristic national police force seeking to quell the turmoil."

One reporter described a British Petroleum drilling site in the eastern foothills of the Andes as "an armed camp, swarming with khaki-clad, rifle-toting guards and surrounded by machine gun emplacements and two rows of flood-lit razor wire." The fortress is necessary in order to quell some of the violence caused by the Marxist guerrillas in the region who are protesting the eradication of coca crops. As it is difficult to halt the coca eradication process, the guerrillas attack the oil sites and pipelines as a demonstration of their dissatisfaction with the government's actions. During the last nine years, "leftist guerrilla squads have dynamited Colombia's main oil pipeline 346 times, spilling more than 1.2 million barrels of crude oil. The guerrillas seek publicity, rural development, and nationalization of the oil industry." The guerrillas also demand increased spending "for social programs in areas where the oil is produced."

Colombia's Environment Minister said that "no one has ever

calculated how much the FARC owes" due to fears that the calculation would encourage the guerrillas to increase their destructive actions. During the past ten years, pipeline attacks by the guerrillas are estimated to have cost Colombia "about $1 billion in lost oil sales." Ministry studies examining the years 1989 through 1991 "found that guerrilla pipeline bombings polluted 375 miles of creeks and rivers and fouled 12,500 acres, ranging from tropical wetlands to Andean watersheds."

## Azerbaijan

On 20 November 1994 a consortium of oil companies signed a contract with the government of Azerbaijan. The consortium, led by British Petroleum, is to invest $8 billion for oil production over a period of 30 years. The consortium believes it can extract up to 4 billion barrels of oil from three wells in the Caspian Sea. However, a problem has developed dealing with the route the oil will take to the world market. In addition, there are many environmental aspects to the issue. They all basically deal with the possibility of damage or destruction of the pipelines. This is due to the fact that this is a politically volatile region of the world.

Before the consortium could finally have the agreement signed, there was a problem that required immediate attention. Their plan is to sell the oil on the world market. As such, the oil must be transported from Baku, the Azerbaijani port of origin to potential world clients by way of the Turkish port of Ceyhan. There were 3 possible routes to be taken. The first involves constructing a pipeline from Baku to the west in neighboring Georgia. From there it would be shipped to Ceyhan. A second would involve constructing a pipeline that would travel south through Armenia into Turkey to Ceyhan. Finally, an existing pipeline could be used by sending the oil north to the Russian port of Novorossiysk, from there it would be shipped to Ceyhan.

The choices to be made are thus influenced by a myriad of details. However, there is one rather important issue which has not been discussed, the environment. The environment serves to be severely damaged if any number of very likely events occur. Firstly, the threat of terrorism on every possible pipeline route is very high. The first route is through Georgia, which has not yet rid itself of the horrors of civil war. Thus there is the possibility of the pipeline being targeted by the combatants. The second route, through Armenia, is the sight of an almost 7 year clash with Azerbaijan over the disputed Nagorno-Karabakh region. The final route through Russia directly traverses the Chechen war zone. In addition, the Russian oil pipelines are in a horrendous state of upkeep, with over 700 spills per year. There is also the likely possibility that the Caspian Sea itself will be polluted by any number of possible mishaps. Finally, if one of the first two oil routes were chosen, oil would have to be shipped to Turkey through the already overcrowded Bosporus Sea channel. Thus animal or fish life, and the very ecosystem itself, could be adversely affected by pipeline or shipping spills.

These options leave much to be desired environmentally. The outcome of an accident such as an oil spill in the area of the Caspian or Black Sea, or in the pipeline system on land would undoubtedly have a high impact on and effect both the composition and scale of the wildlife and its habitat.

## Section-V

# Eating Oil

## Oil Volatility & Food Security

*"And here the rascals are advising, produce bolts and nuts, tire, and drill petrol. They are not producing food. And the so-called government men, they are levying taxes, and they are enjoying. They haven't got to produce food. They are killing animals, eating, and digesting with wine. And then woman. That's all. This is their business. And food price is increasing daily. They don't mind because they will psrint paper, and to the supplier they will give paper. "*
*-Srila Prabhupada*
*(Morning walk, May 25, 1974, Rome)*

Food security is increasingly coming under scanner in recent years as the law of diminishing returns sets in agriculture. How long our fields can support chemical inundation and reckless cultivation methods like monocroping.

Year 2007 saw "great wheat panic" grip the world and year 2008 is witnessing rationing of rice in USA, of all places and food riots in dozens of countries.

Today the food system is completely dependent on cheap crude oil. Virtually all of the processes in the modern food system are dependent upon this finite resource, which is nearing its depletion phase.

Not only is the contemporary food system inherently unsustainable, increasingly, it is damaging the environment.

The systems that produce the world's food supply are heavily dependent on fossil fuels. Vast amounts of oil and gas are used as raw materials and energy in the manufacture of fertilisers and pesticides, and as cheap and readily available energy at all stages of food production: from planting, irrigation, feeding and harvesting, through to processing, distribution and packaging. In addition, fossil fuels are essential in the construction and the repair of equipment and infrastructure needed to facilitate this industry, including farm machinery, processing facilities, storage, ships, trucks and roads. The industrial food supply system is one of the biggest consumers of fossil fuels and one of the greatest producers of greenhouse gases.

Proximity and localisation of food system would be beneficial. The contemporary food system is inherently unsustainable. Ironically, the food industry is at serious risk from global warming caused by these greenhouse gases.

Environmental degradation, water shortages, salination, soil erosion, pests, disease and desertification all pose serious threats to our food supply and are made worse by climate change.

Industrial agriculture and the systems of food supply are also responsible for the erosion of communities throughout the world. This social degradation is compounded by trade rules and policies,

by the profit driven mindset of the industry, and by the lack of knowledge of the faults of the current systems and the possibilities of alternatives. But the globalisation and corporate control that seriously threaten society and the stability of our environment are only possible because cheap energy is used to replace labour and allow the distance between producer and consumer to be extended.

However, this is set to change. We have a very poor understanding of how the extreme fluctuations in the availability and cost of both oil and natural gas will affect the global food supply systems, and how they will be able to adapt to the decreasing availability of energy. In the near future, environmental threats will combine with energy scarcity to cause significant food shortages and sharp increases in prices - at the very least. We are about to enter an era where we will have to once again feed the world with limited use of fossil fuels. But do we have enough time, knowledge, money, energy and political will to make this massive transformation to our food systems when they are already threatened by significant environmental stresses and increasing corporate control?

The modern, commercial agricultural miracle that feeds all of us, and much of the rest of the world, is completely dependent on the flow, processing and distribution of oil, and technology is critical to maintaining that flow.

Oil refined for gasoline and diesel is critical to run the tractors, combines and other farm vehicles and equipment that plant, spray the herbicides and pesticides, and harvest/transport food and seed Food processors rely on the just-in-time (gasoline-based) delivery of fresh or refrigerated food. Food processors rely on the production and delivery of food additives, including vitamins and minerals, emulsifiers, preservatives, colouring agents, etc. Many are oil-based. Delivery in oil-based Food processors rely on the production and delivery of boxes, metal cans, printed

*Food is energy. And it takes energy to get food.*
*-Norman Church*

paper labels, plastic trays, cellophane for microwave/convenience foods, glass jars, plastic and metal lids with sealing compounds.

## Energy, Transport And The Food System

Our food system is energy inefficient. One indicator of the unsustainability of the contemporary food system is the ratio of energy outputs - the energy content of a food product (calories) - to the energy inputs.

The latter is all the energy consumed in producing, processing, packaging and distributing that product. The energy ratio (energy out/energy in) in agriculture has decreased from being close to 100 for traditional pre-industrial societies to less than 1 in most cases in the present food system, as energy inputs, mainly in the form of fossil fuels, have gradually increased.

However, transport energy consumption is also significant, and if included in these ratios would mean that the ratio would decrease further. For example, when iceberg lettuce is imported to the UK from the USA by plane, the energy ratio is only 0.00786. In other words 127 calories of energy (aviation fuel) are needed to transport 1 calorie of lettuce across the Atlantic. If the energy consumed during lettuce cultivation, packaging, refrigeration, distribution in the UK and shopping by car was included, the energy needed would be even higher. Similarly, 97 calories of transport energy are needed to import 1 calorie of asparagus by plane from Chile, and 66 units of energy are consumed when flying 1 unit of carrot energy from South Africa.

Just how energy inefficient the food system is can be seen in the crazy case of the Swedish tomato ketchup. Researchers at the Swedish Institute for Food and Biotechnology analysed the production of tomato ketchup. The study considered the production of inputs to agriculture, tomato cultivation and conversion to tomato paste (in Italy), the processing and packaging of the paste and other ingredients into tomato ketchup in Sweden and the retail and storage of the

*The oil crisis gets louder – listen to it, talk about it, prepare for it – it is out there, the tide is rising and rushing towards us.*
*~James Howard*

final product. All this involved more than 52 transport and process stages.

The aseptic bags used to package the tomato paste were produced in the Netherlands and transported to Italy to be filled, placed in steel barrels, and then moved to Sweden. The five layered, red bottles were either produced in the UK or Sweden with materials from Japan, Italy, Belgium, the USA and Denmark. The polypropylene (PP) screw-cap of the bottle and plug, made from low density polyethylene (LDPE), was produced in Denmark and transported to Sweden. Additionally, LDPE shrink-film and corrugated cardboard were used to distribute the final product. Labels, glue and ink were not included in the analysis.

This example demonstrates the extent to which the food system is now dependent on national and international freight transport. However, there are many other steps involved in the production of this everyday product. These include the transportation associated with: the production and supply of nitrogen, phosphorous and potassium fertilisers; pesticides; processing equipment; and farm machinery. It is likely that other ingredients such as sugar, vinegar, spices and salt were also imported. Most of the processes listed above will also depend on derivatives of fossil fuels. This product is also likely to be purchased in a shopping trip by car.

It is not that this transportation is critical or necessary. In many cases countries import and export similar quantities of the same food products. A recent report has highlighted the instances in which countries import and export large quantities of particular foodstuffs. For example, in

1997, 126 million litres of liquid milk was imported into the UK and, at the same time, 270 million litres of milk was exported from the UK. 23,000 tonnes of milk powder was imported into the UK and 153,000 tonnes exported.

Britain imports 61,400 tonnes of poultry meat a year from the Netherlands and exports 33,100 tonnes to the Netherlands. US imports 240,000 tonnes of pork and 125,000 tonnes of lamb while exporting 195,000 tonnes of pork and 102,000 tonnes of lamb.

This system is unsustainable, illogical, and bizarre and can only exist as long as inexpensive fossil fuels are available and we do not take significant action to reduce carbon dioxide emissions..

*(Ref. Caroline Lucas, 2001. Stopping the Great Food Swap - Relocalising Europe's food supply. Green Party, 2001.)*

In addition, oil is required for cultivation and for pumping water, and on gas for its fertilisers. Also, for every calorie of energy used by agriculture itself, five more are used for processing, storage and distribution.

Most pesticides are petroleum- (oil) based, and all commercial fertilisers are ammonia-based. Ammonia is produced from natural gas. As oil production went up, so did food production.

We are now at a point where the demand for food/oil continues to rise, while our ability to produce it in an affordable fashion is about to drop. Within a few years of Peak Oil occurring, the price of food will skyrocket because the cost of fertiliser will soar, the cost of storing (electricity) and transporting (gasoline) the food that is produced will also soar.

Modern food system is entirely dependent on the thread of technology. Modern, technology-based agriculture produces both food, and seeds for next year's food, on a just-in-time basis. There are precious little reserves of either food or seeds to sustain any protracted interruption. Technology and the incredibly rich tapestry it has made possible has created a false sense of security for so many of us. The thread is flawed; the tapestry is now fragile; famines are possible.

No country is self-sufficient in all respects neither any country is striving in that direction. The implications are grim. For millions, the difference between an energy famine and a biblical famine could well be academic.

Food production is going to be an enormous problem in the Long Emergency. As industrial agriculture fails due to a scarcity of oil-

and gas-based inputs, we will certainly have to grow more of our food closer to where we live, and do it on a smaller scale.

The priority must be the development of local and regional food systems, preferably organically based, in which a large percentage of demand is met within the locality or region. This approach, combined with fair trade, will ensure secure food supplies, minimize fossil fuel consumption and reduce the vulnerability associated with a dependency on food exports (as well as imports). Localising the food system will require significant diversification, research, investment and support that have, so far, not been forthcoming. But it is achievable and we have little choice.

Once we're out of oil, the big consequences will hit. Our planet can only support perhaps a billion people without petroleum-based agriculture. The effects of a failure in production of nitrogenous fertilizers and pesticides and fuel for farming machines will be more pronounced in areas that have been farmed to the point of fatigue.

## Divorce of Oil and Food - An Urgent Necessity

The US food system uses over 10 quadrillion Btu (10,551 quadrillion Joules) of energy each year, as much as France's total annual energy consumption. Growing food accounts for only one-fifth of this. The other four-fifths is used to move, process, package, sell, and store food after it leaves the farm.

Some 28% of energy used in agriculture goes to fertilizer manufacturing, 7% goes to irrigation and 34% is consumed as diesel

*Just like here in this Letchmore Heath there is so much land lying vacant. You produce you own food. Why you are going to London, to the factories? There is no need. This is wrong civilization. Here is land. You produce your food. If you produce your food, there is no need of going hundred miles, fifty miles on your motorcycle or motor to earn your livelihood. Why? There is no need. Then you require petrol. And petrol there is scarcity. Then you require so many parts, so many.. That means you are making the whole thing complicated unnecessarily. Unnecessarily. There is no need.*
*~Srila Prabhupada (Srimad-Bhagavatam Lecture, London, November 25, 1973)*

and gasoline by farm vehicles used to plant, till, and harvest crops. The rest goes to pesticide production, grain drying and facility operations.

The past half-century has witnessed a tripling in world grain production - from 631 million tons in 1950 to 2,029 million tons in 2004. While 40% of the increase is due to population growth raising the demand, the remainder can be attributed to more people eating higher up the food chain (meat). New grain demand has been met primarily by raising land productivity through higher yielding crop varieties in conjunction with more oil-intensive mechanization, irrigation, and fertilizer use, rather than by expanding cropland.

Crop production now relies on fertilizers to replace soil nutrients, and therefore on the oil needed to mine, manufacture, and transport these fertilizers around the world. Rock deposits in the United States, Morocco, China, and Russia meet two-thirds of world phosphate demand, while Canada, Russia, and Belarus account for half of potash mine production. Nitrogen fertilizer production, which relies heavily on natural gas to fuel the conversion of atmospheric nitrogen into reduced forms of nitrogen such as ammonia, is much more widely dispersed.

World fertilizer use has increased dramatically since the 1950s. China is now the top consumer with use rising beyond 40 million tons in 2004. Fertilizer use has leveled off in the United States, staying near 19 million tons per year since 1984. India's use also has stabilized at around 16 million tons per year since 1998. More energy-efficient fertilizer production technology and precision monitoring of soil nutrient needs have cut the amount of energy needed to fertilize crops, but there is still more room for improvement. As oil prices increase and the price of fertilizer rises, there will be a premium on closing the nutrient cycle and replacing synthetic fertilizer with organic waste.

The use of mechanical pumps to irrigate crops has allowed farms to prosper in the middle of the desert. It also has increased farm energy use, allowed larger water withdrawals, and contributed to aquifer depletion worldwide. As water tables drop, ever more powerful pumps must be used, perpetuating and increasing the oil requirements for irrigation. More efficient irrigation systems, such as low-pressure and drip irrigation, and precision soil moisture testing could reduce agricultural water and energy needs. But in many countries, government subsidies keep water artificially cheap and readily available.

Although agriculture is finding ways to use less energy, the amount consumed between the farm gate and the kitchen table continues to rise. While 21% of overall food system energy is used in agricultural production, another 14% goes to food transport, 16% to processing, 7% to packaging, 4% to food retailing, 7% to restaurants and caterers, and 32% to home refrigeration and preparation.

Food today travels farther than ever, with fruits and vegetables in Western industrial countries often logging 2,500-4,000 kilometers from farm to store. Increasingly, open world markets combined with low fuel prices allow the import of fresh produce year round, regardless of season or location. But as food travels farther, energy use soars. Trucking accounts for the majority of food transport, though it is nearly 10 times more energy-intensive than moving goods by rail or barge. Refrigerated jumbo jets - 60 times more energy-intensive than sea transport - constitute a small but growing sector of food transport, helping supply northern hemisphere markets with fresh produce from places like Chile, South Africa, and New Zealand.

Processed foods now make up three-fourths of total world food sales. One pound (0.45 kilograms) of frozen fruits or vegetables requires 825 kilocalories of energy for processing and 559 kilocalories for packaging, plus energy for refrigeration during transport, at the store, and in homes. Processing a one-pound can of fruits or vegetables takes an average 261 kilocalories, and packaging adds 1,006 kilocalories, thanks to the high energy-intensity of mining and manufacturing

## "The End Of Suburbia"
### By Adam Porter
*One of the surprises in the oil world in 2004 was the success of an underground documentary on the perilous state of world energy. "The End of Suburbia" has sold more than a million DVDs and has been aired on TV around the world.*

steel. Processing breakfast cereals requires 7,125 kilocalories per pound - easily five times as much energy as is contained in the cereal itself.

Most fresh produce and minimally processed grains, legumes, and sugars require very little packaging, particularly if bought in bulk. Processed foods, on the other hand, are often individually wrapped, bagged and boxed, or similarly overpackaged. This flashy packaging requires large amounts of energy and raw materials to produce, yet almost all of it ends up in our landfills.

Food retail operations, such as supermarkets and restaurants, require massive amounts of energy for refrigeration and food preparation. The replacement of neighborhood shops by "super" stores means consumers must drive farther to buy their food and rely more heavily on refrigeration to store food between shopping trips. Due to their preference for large contracts and homogenous supply, most grocery chains are reluctant to buy from local or small farms. Instead, food is shipped from distant large-scale farms and distributors - adding again to transport, packaging, and refrigeration energy needs.

Rather than propping up fossil-fuel-intensive, long-distance food systems through oil, irrigation, and transport subsidies, governments could promote sustainable agriculture, locally grown foods, and energy-efficient transportation. Incentives to use environmentally friendly farming methods such as conservation tillage, organic fertilizer application, and integrated pest management could reduce farm energy use significantly. Rebate programs for energy-efficient appliances and machinery for homes, retail establishments, processors, and farms would cut energy use throughout the food system. Legislation to minimize unnecessary packaging and promote recycling would decrease energy use and waste going to landfills.

Direct farmer-to-consumer marketing, such as farmers' markets, bypasses centralized distribution systems, cutting out unnecessary food travel and reducing packaging needs while improving local food security. Farmers' markets are expanding in different parts of the world including India. In US, for example, farmers' markets have grown from 1,755 markets in 1993 to 3,100 in 2002, but still represent only 0.3% of food sales.

The biggest political action individuals take each day is deciding what to buy and eat. Preferentially buying local foods that are in season can cut transport and farm energy use and can improve food safety and security. Buying fewer processed, heavily packaged, and frozen foods can cut energy use and marketing costs, and using smaller refrigerators can slash household electricity bills. Eating lower on the food chain can reduce pressure on land, water, and energy supplies.

Fossil fuel reliance may prove to be the Achilles' heel of the modern food system. Oil supply fluctuations and disruptions could send food prices soaring overnight. Competition and conflict could quickly escalate. Decoupling the food system from the oil industry is key to improving food security.

*(Courtesy: Asia Times, By Danielle Murray, Infoshop News)*

## Blame Game

World leaders are dangerously ignorant of ground realities and are busy blaming each other while the globe steadily slides into a food crisis.

Recently India blamed the US policy of diverting food grains such as corn for producing bio-fuels for the spurt in food grain prices globally. Finance Minister P Chidambaram said in a statement:

"It has been estimated that nearly 20 per cent of corn grown in the United States is diverted for producing bio-fuels. As citizens of one world, we ought to be concerned about the foolishness of growing food and converting it into fuel. Demand for staple food is on the rise, leading to higher prices, but diverting food for fuel has also contributed to increase in food prices. This process (diversion of food grains for bio-fuels) is a sign of the lopsided priorities of certain countries. Prices of maize, rice and wheat, all staple items of food have either doubled or trebled between 2004 and 2008."

The Indian Finance Minister termed the rise in the price of crude oil as "greed" overtaking the common good of the world. The minister also noted that there is no case for raising food prices so high that many poor people could not buy food anymore.

"India imports significant quantities of urea. The price of urea was $175 per metric ton in 2004. By April 2007, it had increased to $288 per MT and in January 2008, it was quoted at $370 per MT," he added.

Noting that there were clear signs of a slowdown in the world economy and signs of rising inflation in many countries of the world, Chidambaram said "global slow down, rising inflation and subdued interest in investment make for a combination that can have only negative consequences for developing countries."

On the other hand, US President George W. Bush has blamed India for the spiraling global food prices.

"There are 350 million people in India who are classified as middle class. That's bigger than America. Their middle class is larger than our entire population," said Bush recently. "And when you start getting wealth, you start demanding better nutrition and better food. And so demand is high, and that causes the price to go up." The US President was endorsing US Secretary of State Condoleezza Rice's earlier comments that the world food crisis was a result of "improvement in the diets of people in India and China."

Stung, Indian government, in a fit of nationalist pique, did not lose time to point out to the US president that his analysis was "completely erroneous."

"George Bush has never been known for his knowledge of economics. And he has just proved once again how comprehensively wrong he is," Jairam Ramesh, Minister of State for Commerce, said,

# Section-VI

# Post-Petroleum Scenario

*To maintain an industrial civilization, it's either oil or nothing.*
*~By Peter Goodchild*

## Chaos That's Coming Our Way

*By Peter Goodchild*

Perhaps the most common response to the peak-oil problem is: "The oil isn't going to disappear overnight. We have a century to prepare." Unfortunately, the fact that the decline in oil is a curve, not a vertical line, makes it difficult to comprehend. What matters is that the serious damage will be done long before we get to those tiny remaining drops a century or so from now. If we look at the forecasts of Petroconsultants Corp., which produces the "bible" of oil data, we can see that in the year 2000 there were five barrels of oil per person per year, but that by 2025 there will only be about two barrels, not five. That's not an "on/off" situation, but at that point the human race should probably wave goodbye to the Oil Economy. The year 2025 is far less than a century from now.

The same statement, "We have a century to prepare," also raises the question: Who is the "we" here? All human beings? A small group of dedicated survivalists? If the answer is the former, then the statement is false: humanity, as a whole, never makes any decisions. The human race, taken in its entirety, simply does not behave in such a sophisticated manner; the human race much prefers ignorance, superstition, cruelty, and intolerance. Robert D. Kaplan's book The Ends of the Earth is one of many texts that elucidate the harsh reality of human nature.

It is not only oil, but in fact the entire economy that has followed a bell curve. The year 1970 was the Peak, the Big Peak of Everything,

> *"This is not good civilization. It will not stay. There will be catastrophe, waiting. Many times it has happened, and it will happen because transgressing the law of nature, or laws of God, is most sinful."*
> ~*Srila Prabhupada (Room Conversation, July 27, 1976, London)*

especially for Americans. Backward or forward on that bell curve, one sees a dirty, noisy, crowded world. Right on that Peak, one sees the Golden Age - Beatlemania, "sex, drugs, and rock 'n' roll," Easy Street. As Dickens might say, "It was the best of times, it was the worst of times."

What about the coming several decades? Of course, a great deal depends on which time period one is discussing: the world of 2100 will be very different from the world of 2020. The question of slow versus fast collapse will also have a big effect on future scenarios. But if we look at tangible events of the last hundred years - the Great Depression of the 1930s, the Soviet collapse of the 1990s, the Argentine collapse of 2001 - two possible conceptions of the future stand out most clearly. These have best been illustrated by novelists (although not with peak oil as the setting) rather than by sociologists.

The first is that of a slow slide into an impoverished police state (George Orwell, 1984). In this scenario, government and banks do not disappear. They are here to curse us forever. We may be poor and living in chaos, but we will live in relentless drudgery, paying taxes and trying to support our mortgages. This is roughly the same scenario as that of the Great Depression of the 1930s - no matter how bad daily life became, the bank was always ready to take away people's houses and farms.

The second is that of a thermonuclear war that throws humanity back into a quasi-medieval world. In the fight for the last drops of oil, civilization is largely destroyed. With Bush's Iran activities, such a scenario is quite plausible. The good news is that governments and banks would be destroyed at the same time. The bad news is that we would be eating a lot of grass soup.

*Supply chains cannot tolerate even 24 hours of disruption. So if you lose your place in the supply chain because of wild behavior you could lose a lot.*
*~Thomas Friedman*

All civilizations grow too large to support themselves, and their leaders have little foresight. These civilizations then collapse and are buried in the mud. The same will happen to modern world, but human shortsightedness prevents us from seeing this.

Roman civilization serves, to a large extent, as a mirror of modern times. The fall of the Roman Empire has been ascribed to various factors, from laziness to lead poisoning. The impoverishment of the soil, and the consequent lack of food, may have played a large part. No doubt it was also a combined military and economic problem: there wasn't enough money to pay for all the soldiers guarding the frontiers. Pestilence may have been another significant factor. Perhaps a more correct answer would actually be a more general one: the empire was too big, and it was poorly led.

The main difference between Modern and previous civilizations, however, is that from now on the cycle of "civilization" cannot be repeated. Oil is not the only mineral that will be in short supply in the 21st century. Industrial civilization has always been dependent on metals, but hematite, for example, is no longer sufficiently common, and mining companies now look for other sources of iron, which can be processed only with modern machinery.

The machines of one century built the machines of the next. The machines of the past - the hammer, anvil, forge, and bellows of the ancient blacksmith - made it possible for later generations to extract the low-grade ores of the present. Very low-grade iron ores can now be worked, but only because there were once better, more accessible ores. This "mechanical evolution" is, of course, liable to collapse: when Rome fell, so did literacy, education, technology. But after many centuries, the Classical world returned. The western world experienced its Renaissance, its rebirth, after the Dark Ages because the natural world was fundamentally unchanged.

In the future, however, after the collapse of the present civilization, the necessary fuels and ores will not be available for such a gradual rebuilding of technology. The loss of both petroleum and accessible ores means that history will no longer be a cycle of empires.

At one point, the money problem will be everything. A few decades later, the money problem will be nothing. Paper money is only a symbol, and it is only valuable as long as people are willing to accept that fiction: without government, without a stock market, and without a currency market, such a symbol cannot endure. Money itself will be useless and will finally be ignored. Tangible possessions and practical skills will become the real wealth.

Most schemes for a post-oil technology are based on the misconception that there will be an infrastructure, similar to that of the present day, which could support such future gadgetry. Modern equipment, however, is dependent on specific methods of manufacture, transportation, maintenance, and repair. In less abstract terms, this means machinery, motorized vehicles, and service depots

*"Yes there was all darkness in New York on the 10th instant and it was not a happy incident. I learn that may people remained in the elevators and in the subway trains for more than seven to eight hours in darkness. I do not read newspapers but there must have been some mishaps also which we may not know. That is the way of material civilization too much depending on machine. At any time the whole thing may collapse and therefore we may not be self complacent depending so much on artificial life. The modern life of civilization depends wholly on electricity and petrol and both of them are artificial for man. You will be surprised to know that I had to take help of the old crude method of lightening by burning some vegetable oil and use the small bowl as lamp to save myself from the extreme darkness. I could not procure any candle from the shop but by the Grace of Krishna one friend Mr. Bill happened to come and he arranged for some fruits and candle. Yes in India we such experience failure of electricity but I was surprised to see the same thing in America. In India the village people say that in Europe (Vilait) also there is ass. And I saw it practically that it is true. Even in such an advanced country there may be possibility such failure but failure in America is more dangerous than that in India. In India they are not so much dependent on electricity but in America the whole activity is dependent on electricity and think on that night it was a great deal of loss to the American nation. Any way the danger is passed but from the papers it is learnt the loss is very great."*
*-Srila Prabhupda (Letter, 13th November, 1965)*

or shops, all of which are generally run by fossil fuels. In addition, one unconsciously assumes the presence of electricity, which energizes the various communication devices, such as telephones and computers; electricity on such a large scale is only possible with fossil fuels.

Without fossil fuels, the most that is possible is a pre-industrial infrastructure, although one must still ignore the fact that the pre-industrial world did not fall from the sky as a prefabricated structure but took uncountable generations of human ingenuity to develop. The next problem is that a pre-industrial blacksmith was adept at making horseshoes, but not at making or repairing solar-energy systems.

Fossil fuels, metals, and electricity are all intricately connected. If we imagine a world without fossil fuels, we must imagine a world without metals or electricity. What we imagine, at that point, is a society far more primitive than the one to which we are accustomed.

We seem to be in a state of delusional thinking and the only thing we're debating at present is how we're going to keep the cars running without oil. How many people nowadays can light a fire without matches or butane lighter from some distant factory?

The skills necessary to get by in a non-industrial society, skills that were still common knowledge a century ago, have been all but lost. Knowledge is critical and currently, there is little knowledge of basic survival skills, and even less knowledge of the scope of the problems that are looming.

Many of the things that we take for granted — food, water, heat, electricity, waste removal, medical care, and police protection — will evaporate as the collapse accelerates. Riots will probably begin as food and water becomes scarce. Governments will attempt to take control of the situation and restore order, but it will become so widespread that it will be impossible. The primary killers will then become disease, starvation, dehydration, and suicide.

Of course once the fossil fuels run out, or become too expensive and/or problematic to extract then there will be no way to rebuild.

There will be no energy source that can power a civilization like this ever again. We will have used it, squandered it and it can not be replaced. Full stop!

There may be pockets of survivors who will be able to harness wind, water and sun using civilized technology for a while, but eventually the machines will wear out. Where do you buy replacement parts, how do you make parts without plastic or wires? How do you refine the metals needed to make circuits and transistors?

Those who know, no longer do; those who do, no longer know. How much knowledge will manage to survive the post collapse period, for the time that comes after when it may become useful again? The problem is that all the technology upon which we have come to depend requires a complete and sophisticated infrastructure to produce and maintain it, and that infrastructure is based on fossil fuels. Take that away, and the rest is all but impossible.

As stated earlier, the ancient Roman world went through very much the same stages as our own. While Rome was a republic, not an empire, the Roman people adhered to the four virtues of prudence, fortitude, temperance, and justice. But the Roman world became bigger and bigger. There were conflicts between the rich and the poor. There was a serious unemployment problem created by the fact that slave labor was replacing that of free men and women. The army became so large that it was hard to find the money to maintain it, and the use of foreign mercenaries created further problems. Farmland became less productive, and more food had to be imported.

*"It is difficult to think about 'how things will play out' when an oil-based global economy loses its cheap energy source. It has never happened before. It will never happen again. I think it quite probable that it will start very slowly, may be so slowly that we may not even see it start.*

*It will take time for civilization to come apart, and the process will be like rolling down a slope, not like falling off a cliff. We will face a future of shortages, economic crises, disintegrating infrastructure, and collapsing public health, probably stretched out over a period of decades." - Norman Church*

The machinery of politics and economics began to break down. The fairly democratic methods of the republic were no longer adequate for a world that stretched from Britain to Egypt, and the emperors took over. After Augustus, however, most of the leaders were both incompetent and corrupt. The Goths sacked Rome in A.D. 410. The Empire was crumbling. The cities and main roads were finally abandoned, since they no longer served a purpose. For the average person, the late Roman world consisted of the village and its surrounding fields.

If we have already established the premise that "the human race faces unsolvable problems," the answer is not to waste further amounts of time and energy in asking whether those problems exist. The best response is to find ways to survive within that problematic world.

To believe that a non-petroleum infrastructure is possible, one would have to imagine, for example, solar-powered machines creating equipment for the production and storage of electricity by means of solar energy. This equipment would then be loaded on to solar-powered trucks, driven to various locations, and installed with other solar-powered devices, and so on. Such a scenario might provide material for a work of science fiction, but not for genuine science. The sun simply does not work that way.

It is not only oil that will soon be gone. Iron ore of the sort that can be processed with primitive equipment is becoming scarce, and only the less-tractable forms will be available when the oil-powered machinery is no longer available - a chicken-and-egg problem. Copper, aluminum, and other metals are also rapidly vanishing.

*"Down one road lies disaster, down the other utter catastrophe. Let us hope we have the wisdom to choose wisely."*
*~Woody Allen*

Metals were useful to mankind only because they could once be found in concentrated pockets in the earth's crust; now they are irretrievably scattered among the world's garbage dumps.

The infrastructure will no longer be in place: oil, electricity, and asphalt roads. Partly for that reason, the social structure will also no longer be in place: intricate division of labor, large-scale government, and high-level education. Without the infrastructure and the social structure, it will be impossible to produce the familiar goods of industrial society.

Then prices will rise and demand will fall. The rich will outbid the poor for available supplies. The system will initially appear to rebalance. The dash for gas will become more frenzied. People will realize nuclear power stations take up to ten years to build. People will also realize wind, waves, solar and other renewables are all pretty marginal and take a lot of energy to construct. There will be a dash for more fuel-efficient vehicles and equipment. The poor will not be able to afford the investment or the fuel.

Exploration and exploitation of oil and gas will become completely frenzied. More and more countries will decide to reserve oil and later gas supplies for their own people. Air quality will be ignored as coal production and consumption expand once more. Once the decline really gets under way, liquids production will fall relentlessly by 5%/ year. Energy prices will rise remorselessly. Inflation will become endemic. Resource conflicts will break out.

## Survival By Localization

The most basic principle of post-oil survival is that one has to start thinking in terms of a smaller radius of activity. The globalized economy has to be replaced by the localized economy.

There would only be three practical methods of travel: on foot, in a non-motorized boat, or on horsebasic. In Asia, bull would reign. One's speed by any of these three methods will be about the same: 25 miles per day, if one is in good shape. Even where paved roads are usable, bicycles would be hard to repair without the industrial infrastructure to provide the spare parts and the servicing.

Those who live in the country will be better prepared than those who live in the city. A city is a place that consumes a great deal and produces little, at least in terms of essentials. A city without incoming food or water collapses rapidly, whereas a small community closely tied to the natural environment can more easily adjust to technological and economic change.

## Some Case Studies

What will be the effects of peak oil on the ordinary citizen and how will we see it coming? On this page we see the likely effects of the future by examining the past.

# 2000 Fuel Protests

It was the worst oil crisis in Europe in a couple of decades. But this was not like the energy crisis of the past. This was fundamentally a political crisis. There was no shortage of oil in Britain or indeed on the world market. There was no war that was disrupting supply from the Middle East. This was not OPEC action, oil cartel disrupting supply for political reasons. What we actually had here is a good study in how to bring an advanced industrialized economy to its knees. A very relatively small band of users of gasoline, truck drivers, farmers, independents mostly, not heavily unionized, they targeted refineries, the choke points in the modern industrial economy. Because of a very poor political response, first in France where the government buckled, the government gave in to their demands, and then in Britain and elsewhere in Europe, the chaos caught on like wildfire.

Many countries were almost brought to a halt. It was deeply alarming.

## Blackouts

The blackouts that hit the eastern USA and Canada in August 2003, and the lesser failure that hit London's Underground system shortly afterwards shows how totally dependent we are on electric power and the dramatic effects that its absence causes. Standard and Underground trains came to a halt, trapping people within; lifts stopped between floors; street lamps failed; people poured from the buildings, all increasing the risk of accidents. People were unable to communicate because they had switched from land phones to mobiles. The pressure on the emergency services was immense.

Imagine the effects when it was the whole country, not just a few cities that is hit. And if that blackout lasts for days, with more occurring in following weeks.

Blackouts can put the very substance of civilization at risk. Prolonged blackouts can take us back to the dark ages – in more respects than one.

## Lawlessness

The hurricane and flooding that struck New Orleans and the surrounding areas of the southern United States in August and September 2005 showed just how quickly the most organized society can break down. Although the affected area was huge, most of the

*Dr. Singh : Now there is a big petrol problem, a shortage of oil.*
*Srila Prabhupada : Yes. The scientists have created it. They have built a civilization that is dependent on oil. This is against nature's law, and therefore there is now an oil shortage. When the petrol supply dwindles away, what will these rascal scientists do? They are powerless to do anything about it. By nature's law, winter is coming. Scientists cannot stop it and turn it into summer. They wrongly think they are in control of nature. (BTG 1979)*

USA was unaffected and therefore still had a fully working infrastructure, security and government. Yet the inhabitants of New Orleans were left to fend for themselves for a week, resulting in looting (both for gain and survival), widespread crime and death. Imagine the results if it was a whole country affected, and if the security and emergency forces were unable to help because of oil shortages. It is an excellent example of what can happen when both security and the basic necessities of life are removed.

# Section-VII

# Viability Of Alternatives

First of all, let us compare the relative energy derived from different types of fuel. For instance, straw only yields around 15 megajoules per kilogram (MJ/kg), wood yields 20 MJ/kg, coal yields 20-25 MJ/kg, while fossil fuels (oil, gasoline, jet fuel) yield around 42 MJ/kg. The corresponding energy densities are 1000 watt per square meter for wood, 5,000 watts per square meter for coal, and 40,000-50,000 watts per square meter for fossil fuels. The energy density of fossil fuels is unparalleled and it is the reason industrial societies have become so dependent on them.

Before we run out of fossil fuels, we will run out of the cheapest and most easily accessible fuel. Soon, it will not be economical to continue to rely on fossil fuels. However, alternative energy sources are not without their own short comings. For instance, renewable sources require huge areas to gather the energy. To heat a city the size of, lets say, Montreal with fuel wood would require hundreds of thousands of hectares of forest lands. Wind farms would require massive areas under turbines and such wind farms may cause considerable change to wind and atmospheric dynamics, upsetting ecological systems. Furthermore, were it not for massive subsidies, wind power would not be competitive with other energy sources. Hydropower can only capture about 30-40% of the total potential energy in a given basin, and most basins in Europe and North America are saturated. Hydropower has been almost entirely captured in Europe and North America.

*"One thing is clear: the era of easy oil is over. What we all do next will determine how well we meet the energy needs of the entire world in this century and beyond." – David J. O'Reilly, Chairman & CEO of Chevron, July 2005.*

## Alternative Source of Power

### Coal

There are still an estimated 909 billion tonnes of proven coal reserves worldwide, enough to last at least 155 years. But coal is a fossil fuel and a dirty energy source that will only add to global warming.

### Natural Gas

The natural gas fields in Siberia, Alaska and the Middle East should last 20 years longer than the world's oil reserves but, although cleaner than oil, natural gas is still a fossil fuel that emits pollutants. It is also expensive to extract and transport as it has to be liquefied.

### Hydrogen Fuel Cells

Hydrogen fuel cells would provide us with a permanent, renewable, clean energy source as they combine hydrogen and oxygen chemically to produce electricity, water and heat. The difficulty, however, is that there is not enough hydrogen to go round and the few clean ways of producing it are expensive.

### Biofuels

Ethanol from corn and maize has become a popular alternative to oil. However, studies suggest ethanol production has a negative effect on energy investment and the environment because of the space required to grow what we need.

Biofuel refers to fuels derived from recently living organisms, today mostly in the form of ethanol from plants such as sugar cane, soybeans, and oil palm. Biofuels often use more energy to produce than they contribute. Scientists hope that biofuels can replace much gasoline used today. The use of biofuels has led to horrific consequences for

*Now they are crying, "Where is the petrol? Where is the petrol?" So if nature does not supply petrol, then all these horseless carriage will be pieces of tin. That's all.  – Srila Prabhupada*

the people of the world and the environment. This will continue to mean that the growing of crops for fuel, mostly for export to Europe, Japan and the US, is being done on large-scale plantations in the third world. Ancient forests are being cut down, threatening extinction for many species. Reduction of greenhouse gases is lost when carbon-capturing forests are cut down. In Malaysia, the production of palm oil for biodiesel is a major industry. The development of oil-palm plantations was responsible for an estimated 87% of deforestation. In Sumatra and Borneo, 4 million hectares of forest have been converted to palm farms. Now a further 6 million hectares are scheduled for clearance in Malaysia, and 16.5 million in Indonesia. Thousands of indigenous people have been evicted from their lands, and some 500 Indonesians have been tortured when they tried to resist. The forest fires which every so often smother the region in smog are mostly started by the palm growers. Hundreds of thousands of small-scale peasant farmers are being displaced by soybeans expansion. Many more stand to lose their land under the biofuels stampede. The expanding cropland planted to yellow corn for ethanol has reduced the supply of white corn for tortillas in Mexico, sending prices up 400%.

For investors in alternatives to oil and gas, the driving force has been the belief that whoever develops the next great energy sources will enjoy the spoils that will make the gains from creating the next Amazon.com or Google seem puny. In the development of biofuels this means that they do not pay attention to long-term effects. The economy is broken up into competing units of capitalist control and ownership over the means of production. And each unit is fundamentally concerned with itself and its expansion and its profit. The economy, the constructed and natural environment, and society cannot be dealt with as a social whole under capitalism.

*The use of solar energy has not been opened up because the oil industry does not own the sun.*
*~Ralph Nader*

## Massive Diversion of U.S. Grain to Fuel Cars

Corn prices have doubled over the last year, wheat futures are at their highest level in 10 years, and rice prices are rising. The use of corn as the feedstock for fuel ethanol is creating consequences throughout the global food chain. Food prices are rising in China, India, and the US, housing 40% of the world's people. Vast quantities of corn are consumed indirectly in meat, milk, and eggs in both China and the US. In China, pork prices were up 20% above a year earlier, eggs were up 16%. In India, the food price index in 2007 was 10% higher than a year earlier. The price of wheat has jumped 11% and this is only the beginning. As more and more fuel ethanol distilleries are built, world grain prices are starting to move up toward their oil-equivalent value. In this new economy, if the fuel value of grain exceeds its food value, the market will move it into the energy economy. With 80 or so ethanol distilleries under construction, nearly a third of the 2008 grain harvest will be going to ethanol. Since the United States is the leading exporter of grain, what happens to the U.S. grain crop affects the entire world. The world's breadbasket is fast becoming the U.S. fuel tank. The UN lists 34 countries as needing emergency food assistance. Food aid programs have fixed budgets. Protests in response to rising food prices could lead to political instability that would add to the list of failed and failing states. President Bush has set a production goal for 2017 of 35 billion gallons of alternative fuels. The risk is that millions of those on the lower rungs of the global economic ladder will start falling off as higher food prices drop their consumption below the survival level.

In 2007, 18,000 children were dying every day from hunger and malnutrition.

Any fuel derived from any plant requires the availability of fertile topsoil, water, increasingly complex and expensive fertilizers, pesticides, and an appropriate climate. The first, second, and last of these are presently threatened. The use of biofuels is therefore a short term solution with disastrous longterm consequences. Biofuels will directly compete with food. This will increase intra- and international tensions and consequently conflicts. The production of biofuels will increase deforestation and contribute to loss of biodiversity. Of course, farmers, particularly industrial agricultural enterprises, will derive short term profits and therefore like it.

Plants need some 30 different soil elements to grow. If any of them becomes depleted, growth will decrease and ultimately cease. Biofuel production will accelerate the speed with which the finite amount of fertile global topsoil is degraded.

## Renewable energy

As already explained in previous sections, oil-dependent nations are turning to renewable energy sources such as hydroelectric, solar and wind power to provide an alternative to oil but the likelihood of renewable sources providing enough energy is slim.

These sources: solar, wind, nuclear, tidal, etc. are not as energy dense, portable, or as readily usable as fossil fuels. History tells us that complete development of new energy sources (coal and oil in the past) takes about a century.

## Nuclear

Fears of the world's uranium supply running out have been allayed by improved reactors and the possibility of using thorium as a nuclear fuel. But an increase in the number of reactors across the globe would increase the chance of a disaster and the risk of dangerous substances getting into the hands of terrorists.

## Can Nuclear Energy Replace Oil?

We can draw an economic model to calculate whether or not nuclear could replace oil. *(By G.R. Morton, http://home.entouch.net/dmd/nuke.htm)*

6.29 bbl (barrels) = 10.9 megawatt-hour

1 bbl = 1.73 megawatt-hour

700megawatt = 404.6242775 bbl/hour (http://www.cameco.com/investor_relations/annual/2001/glossary/index.php)

There are 8760 hours per year so a 700 megawatt plant produces:

6,132,000 megawatt-hour = 3,544,508 bbl/yr

We produce around 30 billion barrels of oil per year. So for the world to replace this we need:

8463.796477 700 MW plants

$1,250,000,000 per plant

$10.5 trillion investment

The US GDP is about $10 trillion. This represents about 1/3 of the global domestic product. Given that they take about 4 years to build the investment would mean a 10% tax on everyone and every corporation--and that would mean that the people would pay far more than 10% of their personal income. Corporations don't pay taxes, they pass them on to consumers.

And these costs don't include the cost of getting rid of the nuclear waste. This is only for building the things. So looks like we will not be able to replace oil with nuclear.

There is continuing debate over whether a suitable energy alternative might be found to replace the energy from oil as it runs out, but there is certainly no compelling evidence that a comparable substitute will be found.

It is difficult to think about 'how things will play out' when an oil-based global economy loses its cheap energy source. It has never happened before. It will never happen again.

Incidentally, "alternative energy" doesn't work. As John Gever et al. explain in "Beyond Oil", it is physically impossible to use windmills etc. to produce the same amount of energy that we are now getting from thirty billion barrels of oil. "Alternative energy" will never be able to produce more than the tiniest fraction of that amount.

The energy budget must always be positive and output must exceed input. Too much tends to be expected of renewable energy generators today, because the contribution of fossil fuels to the input side is poorly understood. For example, a wind turbine is not successful as a renewable generator unless another similar one can be constructed from its raw materials using only the energy that the first one generates in its lifetime, and still show a worthwhile budget surplus.

Or, if corn is grown to produce bioethanol, the energy input to ploughing, sowing, fertilizing, weeding, harvesting and processing the crop must come from the previous year's bioethanol production. Input must also include, proportionately, mining and processing the raw materials and building the machines that do the work, as well as supporting their human operators.

There is nothing that can replace cheap oil for price, ease of storage, ease of transportation and sheer volumes in the timeframe we need.

### EROEI and EPR

An important element in comparing fossil fuels with other forms of energy generation goes by the unfortunately unmemorable acronym of EROEI – "energy returned on energy invested". An alternative version of this is the EPR – Energy Profit Ratio. To produce any energy, whether it is pumping oil out of the ground, or building and operating a wind turbine, you need to use some energy in the process. If the energy returned is less than the energy you produce, it is generally not worth producing it.

As a simple example, imagine a (very small) car whose fuel tank holds 1 liter of petrol. The car's fuel efficiency is 20 km per liter. If the nearest petrol station is 5 km away, fine – you wait until the tank is quarter full then drive there to refill (positive EROEI). If it is 10 km away, you have gained nothing (and lost money) – by the time you have refuelled and driven home, you only have enough fuel left to return to the station to fill up again. And if the station is 15 km away, once you have filled up and reached home, you would not have enough left to get back to refuel again. You would be better off staying at home and simply using up the existing petrol for other journeys (negative EROEI).

The EROEI is calculated by taking the energy content of your energy (in whatever units you wish) and subtracting the energy used in producing the energy. The result will be a number either negative, positive or zero. The higher the number, the better.

The only time when negative EROEI can be worthwhile is if the energy produced is in a more useful form than the energy used. For example, oil can be used not only for energy generation but to make petrochemicals whereas wind-generated electricity cannot. So it could be more worthwhile using some wind-electricity to pump oil-energy out, even if the EROEI is negative. Using the car analogy above, if the journey to the 15 km petrol station was also used to deliver some goods to sell, you would gain elsewhere even if you lost out on the petrol.

*"As soon as you make misuse, the supply will be stopped. After all, the supply is not in your control. You cannot manufacture all these things. You can kill thousands of cows daily, but you cannot generate even one ant. And you are very much proud of your science. You see. Just produce one ant in the laboratory, moving, with independence. And you are killing so many animals? Why? So how long this will go on? Everything will be stopped.*
*Just like a child. Mother is giving good, nice foodstuff, and he's spoiling. So what the mother will do? "All right. From tomorrow you'll not get." That is natural."*
*–Srila Prabhupada (Lecture on Bhagavad-gita, Los Angeles, December 27, 1968)*

# Section-VIII

# The Solution

*For centuries, all our spiritual traditions have told us to avoid
focusing on accumulation of material goods. The familiar "Sermon
on the Mount," based on the theme from Matthew 16:26 "For
what doth it profit a man, if he gain the whole world, and suffer the
loss of his own soul?" is rarely heard today. In the last half of the
20th century, particularly the last 25 years, the West has rejected
these traditions and adopted the religion of growth economics, the
fundamental principle being that if everyone pursues his or her own
self interest, then society will benefit.*
*~George Monbiot*

# Solution -1

## Changing Philosophy Of Life

Discarding Polluted Philosophies

And

Adopting   Healthy   Philosophies

*"Without a global revolution in the sphere of human
consciousness a more human society will not emerge."
-Vaclav Havel
(Former president of Czechoslovakia)*

## Discarding Polluted Philosophies Of Life

Root of petroleum crisis, or all other man-made crises for that matter, lies in polluted philosophies of life. Mind is the place of first creation. All crises in this world, including that of resource and environment are the direct outcome of our polluted ideas about life and World. These polluted ideas lead to polluted activities which in turn lead to different types of perplexities.

## From Overconsumption To Moderation
### Doing Away With Material Extravagance And Inefficacies

Overconsumption lies at the heart of petroleum crisis. Overpopulation is not a problem, overconsumption is.

Over-consumption is a concept referring to situations where per capita consumption is so high that even in spite of a moderate population density, sustainability is not achieved. For example, the People's Republic of China has an area comparable to that of the United States of America. China's population density is 4.7 times higher than that of the USA, but its per capita energy consumption is nine times lower than that of the USA, so that in spite of its larger population, China uses only half the amount of energy consumed by the USA.

Americans constitute less than 5% of the world's population, but produce 25% of the world's $CO_2$, consume 25% of world's resources, including 26% of the world's energy, although having only 3% of the world's known oil reserves, and generate roughly 30% of world's waste. America's impact on the environment is at least 250 times

*"Society is created by and composed of humans. Society as such is thus influenced and transformed in no small way by how humans comprehend and understand society."*

greater than a Sub-Saharan African. It was all fine as long as Americans lived that way. But the problem is that the whole World is now trying to follow in their footsteps.

So far, we have only one usable planet. The astronauts are trying to discover if there are any planets out there where we can go and live. This has not produced any results and there is no future hope in this regard either. This leaves us to face the fact that the 6.5 billion of us are dependent on the natural resources that exist here on this planet. Unfortunately, we are using those resources in an unsustainable way.

850 million humans go hungry today out of which 220 million are children. 1 in 5 humans have no access to clean drinking water. By 2050, 85% of all humans will be living in developing countries. One third of the world's visible land is affected by desertification, the degradation of productive but fragile lands which have insufficient rainfall and has been damaged by unsustainable development. During the next 100 years global temperatures will rise by 2 to 6 degrees C, resulting in coastal flooding and an increase in droughts.

More than one century of industrial development, economic growth and intensive exploitation of nature has led us to a world where we can travel cheaply to anywhere in the world, import food, clothes and materials from any country, yet we are slowly destroying the very earth which keeps us alive. Despite increase in automation, we are working harder and neglecting our lives, both internal and external, to fulfil the demands of industry. The brave new world promised to us by technology has not arrived. Instead it has led to an increased gap between rich and poor within most countries and

*isavasyam idam sarvam*
*yat kinca jagatyam jagat*
*tena tyaktena bhunjitha*
*ma grdhah kasya svid dhanam*
*Everything animate or inanimate that is within the universe is controlled and owned by the Lord. One should therefore accept only those things necessary for himself, which are set aside as his quota, and one should not accept other things, knowing well to whom they belong.*
*-Sri Isopanisad*

between countries. Cultures once unique and distinct are slowly merging into variations of Western popular culture. Languages and species are dying out.

The outlook is gloomy! We have a very strong responsibility to the future which we will create. If we wish to improve the situation, we will thus have to modify our ways of life and our manner of consuming because all resources crunch is due to overindulgence. The uneven distribution of food is not due to shortages but due unbridled greed. Too much land is being exploited for cash crops, junk foods, exports, tobacco, alcohol. Agribusiness is destroying small farms, food prices are soaring, and soil and forests are disappearing fast.

Only a spiritual paradigm can help us check the imbalance of values in life and achieve real unity and peace in the world.

## I Am The Lord Of All I Survey

### Man Is The Master Of The Universe

This misguided thinking also lies at the root of the petroleum crisis. It's hard for most people to identify this ultimate cause of the crisis, because it happens to be spiritual in nature.

This faulty paradigm of life emboldens mankind to defy nature and its complex laws. Present crisis stems from a crisis in consciousness and a spiritual vision of the universe is the key to bring our planet to a more healthy condition.

The overconsuming, overdeveloped lifestyles and industries of the 'minority world' have depended upon the military and economic oppression of labour and ecosystems of the 'majority world'. For everyone on this planet to 'enjoy' the materialistic lifestyle of the average American or Australian, we would need five to six Earths in order to supply the necessary raw materials, handling of consumer and industrial wastes, and life-sustaining services such as clean air and water.

Now the present crisis of petroleum is precisely stemming from this. Asians are waking up to American dream and there is simply not enough oil to facilitate this. Hence the crunch.

Greed And Affluence Is Progress, Simplicity Is Antiquated

Mahatma Gandhi argued that "the world has enough for everyone's need, but not for everyone's greed." Oil crisis is certainly one of the most terrible offshoots of globalization and rampant consumerism.

This idea has not only finished off oil but also our environment. Forget about oil, its time to worry about water. Take the example of Bangalore, India. It is one of the cities that have been terribly affected by the indiscriminate cutting down of trees and sealing in of lakes in the name of growth and industrialization. Until about ten years ago, Bangalore was known as an air-conditioned city, without any summer practically. But in the past few years, everybody is buying air-conditioners as the temperatures have soared.

In 1961, the city boasted of 262 lakes, considered to be a part of Bangalore's green belt. Presently only 61 of them are alive while the remaining have made way for concrete structures. Faced with acute water crisis, the city is spending 1800 crore rupees to recycle sewage. What goes out comes right back in!

## Not In My Backyard Syndrome (Nimby)

Also if we look from another angle, petroleum crisis originates from a degraded self-centered mentality of me and mine. On individual, social and national level, more than ever before, we have become more self-centered. I should be fine and rest of the world can go to hell. President Bush is more concerned about the cough of his dog than the death of a million Iraqis. This phenomenal selfishness, apathy and unconcern is responsible for uneven distribution of resources and irresponsible exploitation of the same.

"The richest billion people in the world have created a form of civilization so acquisitive and profligate that the planet is in danger," says Alan Durning of the Worldwatch Institute. "The life-style of this top echelon-the car drivers, beef eaters, soda drinkers, and throwaway consumers-constitutes an ecological threat unmatched in severity."

When you think of the world as a system, you understand that air pollution from North America affects air quality in Asia, and that pesticides sprayed in Argentina could harm fish stocks off the coast of Australia. Therefore every individual, community or nation has to realize that its destiny is inextricably connected to the rest of

*The scientists are very much busy that the source of supply is being decreased. Just like petroleum. Petroleum, gas, that is diminishing. Now, whole modern materialistic civilization is depending on the motorcars and aeroplanes, transportation. So if the petroleum supply is stopped, then what will be the condition of the society?*

*Formerly there was no need of going to see a friend thirty miles away, because every friend was within the village. Now, because we have got motorcar, we create friendship with a man who lives fifty miles away. We accept a job fifty miles away. In Hawaii our Gaurasundara was going to attend office fifty miles off. In big, big cities like New York, Calcutta, we have seen people coming to attend their offices from hundred miles off. I have seen... I have seen in England. Many workers or gentlemen, they are coming from Glasgow to London for working, by aeroplane.*

*-Srila Prabhupada (Srimad-Bhagavatam Leture, Los Angeles, September 16, 1972)*

the world. If we are to float, we will float together, if we were to sink, we will sink together. Nature is one complex system and no nation can isolate itself from it.

We can conclude by saying that simplifying our life style and reducing consumption of petrol is one sure way of tackling oil crisis.

# Solution -2

## Do Away With
## Materialistic World View And Godless Cosmology

*It's not an if. We're going to have to change. Oil is simply going to be gone.*
*~Dennis Weaver*

## Godless Science and Worldview

Resource crisis is the direct outcome of over-consumption which in turn is the consequence of our excessive material pursuits. Up until a few centuries ago, people all over the world, including in progressive Europe, led a simple God-centred life wherein the main focus was spiritual elevation. Material progress was assigned a secondary place. This life style was in harmony with the available natural resources and it did not leave any destructive footprints on the natural world.

For the first time in human history, in preindustrial Europe of the Middle Ages, there was paradigm shift as far as purpose and destination of human life was concerned. This was also the turning point for the world ecology. The idea contained in the phrase "What profiteth a man if he gains the whole world and loses his own soul" began to fade away. As the human society began to move away from God-centredness, it became more and more callous to nature and life in general.

Faith in God necessitated faith in the 'other' world and therefore the earthly years were not taken to be-all and end-all. Some kind of ascetism was also associated with every religious practice and all this counted favorably for the natural resources and environment.

Indian civilization too shared the vision of a divine purpose in life. Srimad-Bhagavatam, a five thousand year old treatise, contains this passage about the ultimate purpose of human life: "The supreme occupation [dharma] for all humanity is that by which men can attain to loving devotional service unto the transcendental Lord."(SB 1.2.6)

Seed of modernization fructified in Europe and 'American dream' was the mature fruit of it. This fruit of American way of life has become the goal of the entire world. Unfortunately this way of life is not at all sustainable.

Materialism inflicts a heavy toll on limited resources like oil because materialism is the idea that everything is either made only of matter or is ultimately dependent upon matter for its existence and nature. It is possible for a philosophy to be materialistic and still accord spirit a (secondary or dependent) place, but most forms of materialism tend to reject the existence of spirit or anything non-physical.

Because materialists only accept the existence or primacy of material things, they also only accept the existence or primacy of material explanations for events. Whatever happens in the world, it must be explained and explainable by reference to matter. Materialism thus tends towards determinism: because there are material causes for every event, then every event follows necessarily from its causes.

Materialism is closely associated and aligned with the natural sciences. Modern science involves the study of the material world around us, learning about material events, and theorizing about their material causes. Scientists are materialists in that they only study the material world, although they may personally believe in non-material entities. Science in the past has tried to incorporate vitalist ideas and the supernatural, but those efforts failed and have since been discarded.

Atheists are usually materialists of some sort, rejecting the idea that there exists anything independent of the workings of matter and energy. Materialism often entails atheism unless a person believes in a purely physical god.

## A Western History Of Materialism

Root of petroleum crisis lies in advent of materialism and its important to understand the history of materialism.

We find that many early scientists were believers in special creation. Pasteur, Mendel, and Faraday are examples. We should examine the lives of these early scientists and their lives to recapture their spirit and curiosity about God's creation. About 96 percent of innovators from the mid-1500s to 1700 were believers. And the great majority of those, about 60 percent, led devout lives.

Newton saw God as the master creator whose existence could not be denied in the face of the grandeur of all creation. Although the laws of motion and universal gravitation became Newton's best-known discoveries, he warned against using them to view the universe as a mere machine, as if akin to a great clock. He said, "Gravity explains the motions of the planets, but it cannot explain who set the planets in motion. God governs all things and knows all that is or can be done." *(Tiner, J.H., 1975. Isaac Newton: Inventor, Scientist and Teacher)*. Copernicus wrote that their orbits were an illustration of the "divine work of the Great and Noble Creator."

Thus even as Europe entered the age of scientific discovery, many of the early scientists retained deep and profound conceptions of God as the ultimate controller and designer of the universe.

Renaissance and Enlightenment eras in Europe saw the growth of science but God was gradually moved out of the scene. What Newton cautioned against precisely came to be practiced. He had warned against using science to view the universe as a mere machine, governed not by divine arrangement but by precise, mathematically expressed physical laws. Thus was laid the foundation for materialistic modern science and Godless worldview. Descartes in 1637 introduced the idea of reductionism. Descartes argued the world was like a machine, its pieces like clockwork mechanisms, and that the machine could be understood by taking its pieces apart, studying them, and then putting them back together to see the larger picture. Reductionist thinking and methods are the basis for many of the well-developed areas of modern science, including much of physics, chemistry and cellular biology. Thus reductionism threw God out of the picture and reduced the universe, including all human experience, to measurable and predictable states or actions of matter (ultimately subatomic particles) and material forces (such as electromagnetism and gravity).

## Oil Crisis Founded on Disregard for Nature and Natural Living

The scientific institution thus began to lose concern for God and concern for nature as well. The nature that was seen as a handiwork of God and to be revered, cooperated and preserved suddenly became a subject to be dominated and made a slave. Sir Francis Bacon (1561-1626), one of the main founders of the modern scientific method, viewed nature as a mysterious, wild woman, something to be exploited and plundered.

Francis Bacon set forth the empirical method. Bacon was the first to formulate a clear theory of the inductive procedure - to make experiments and to draw general conclusions from them, to be tested in further experiments - and he became extremely influential by vigorously advocating the new method. He boldly attacked traditional schools of thought and developed a veritable passion for scientific experimentation.

The "Baconian spirit" profoundly changed the nature and purpose of the scientific quest. From the time of the ancients the goals of science had been wisdom, understanding the natural order and living in harmony with it. Since Bacon, the goal of science has been knowledge that can be used to dominate and control nature, and today both science and technology are used predominantly for purposes that are profoundly antiecological.

Nature, in his view, had to be "hounded in her wanderings", "bound into service", and made a "slave". She was to be "put in constraint", and the aim of the scientist was to "torture nature's secrets from her".

The ancient concept of the earth as a nurturing mother was radically transformed in Bacon's writings, and it disappeared completely as the Scientific Revolution proceeded to replace the organic view of nature with the metaphor of the world as the machine.

> *The Western civilization is a nasty civilization, artificially increasing the necessities of life. For example, take the electric light. The electric light requires a generator, and to run the generator you need petroleum. As soon as the petroleum supply is stopped, everything will stop. But to get petroleum you have to painstakingly search it out and bore deep into the earth, sometimes in the middle of the ocean. This is ugra-karma, horrible work. The same purpose can be served by growing some castor seeds, pressing out the oil, and putting the oil into a pot with a wick. We admit that you have improved the lighting system with electricity, but to improve from the castor-oil lamp to the electric lamp you have to work very hard. You have to go to the middle of the ocean and drill and then draw out the petroleum, and in this way the real goal of your life is missed.*
> *- Srila Prabhupada (Room Conversation, June 24, 1976, New Vrindaban)*

In 1626 Bacon wrote a utopian novel, The New Atlantis. It depicts a mythical land, Bensalem, to which he sailed, that was located somewhere off the western coast of the continent of America. He recounts the description by one of its wise men, of its system of experimentation, and of its method of recognition for inventions and inventors. The best and brightest of Bensalem's citizens attend a college called Salomon's House, in which scientific experiments are conducted in Baconian method in order to understand and conquer nature, and to apply the collected knowledge to the betterment of society. In this novel, Bacon listed some of the inventions he could foresee: "The prolongation of life ... means to convey sound in trunks and pipes in strange lines and distances ... flying in the air ... ships and boats for going under water." Also in the list: "instruments of destruction as of war and poison" and "engines of war, stronger and more violent, exceeding our greatest cannons."

Thanks to these developments, it did not take science and technology long to begin playing a primary role in people's lives. Spiritual quest came to be assigned a secondary place. The consequences were devastating. Individuals and cultures were stripped of inner meaning and the external world (including the global ecology) was rendered into a set of things, mere resources. Consequently the world of modernity was built on an illusion: the illusion that only half of reality mattered: the external, objective, measurable part. The cry 'no more myths' led to the abandonment of any possibility of further development and to the 'disenchantment' of self and the world. Historian Lewis Mumford said, "Whatever their adhesion to the outward ceremonies of the Church . . . more and more people began to act as if their happiness, their prosperity, their salvation were to be achieved on the earth alone, by means they themselves would if possible command."

Eighteenth and nineteenth centuries further saw the degradation of human spirit and further rise of materialism. Thus even greater thrust was laid into building the machines Bacon envisioned, vastly increasing human ability to exploit the earth's resources. Thus began the industrial revolution, which is responsible for the energy crisis we face now.

## The Onset of Mining Industry

The story then begins with mining which had to face considerable resistance in the beginning. Not only medieval Europe was skeptical about the activity of mining but way back even Greek and Roman civilizations expressed their reservations on excavating earth for resources. Prophetic words of Pliny (A.D. 23-79), who wrote them in his work called 'Natural History', deserve a mention.

"For it is upon her surface, in fact, that she has presented us with these substances, equally with the cereals, bounteous and ever ready, as she is, in supplying us with all things for our benefit! It is what is concealed from our view, what is sunk far beneath her surface, objects, in fact, of no rapid formation, that urge us to our ruin, that send us to the very depths of hell. As the mind ranges in vague speculation, let us only consider, proceeding through all ages, as these operations are, when will be the end of thus exhausting the earth, and to what point will avarice finally penetrate! How innocent, how happy, how truly delightful even would life be, if we were to desire nothing but what is to be found upon the face of the earth; in a word, nothing but what is provided ready to our hands!"

Thus the medieval opponents of large-scale mining had a surprisingly prophetic view of its negative environmental effects.

The famous German scholar of mining and metallurgy, Georgius Agricola, stated in his 1556 treatise "De Re Metallica" that :

"... the strongest argument of the detractors is that the **fields are devastated** by mining operations ... Also they argue that the **woods and groves are cut down,** for there is need of an endless amount of wood for timbers, machines, and the smelting of metals. And when the woods and groves are felled, **then are exterminated the beasts and birds,** very many of which furnish a pleasant and agreeable food for man. Further, **when the ores are washed, the water which has been used poisons the brooks and streams,** and either destroys the fish or drives them away. Therefore the inhabitants of these regions, on account of the devastation of their fields, woods, groves, brooks and rivers, find great difficulty in procuring the necessaries of life ... **Thus it is said, it is clear to all that there is greater detriment from mining than the value of the metals which the mining produces.**" *(Emphasis added)*

Next toll of declining God-centered world view and rising materialism was local and self-sufficient agricultural system. Between 1500 and 1700, market economy emerged signalling the demise of subsistence farming. Industrialization of agriculture set in motion a process that is even now destroying traditional village economies and the environment.

In the village communities of many areas of medieval Europe, land was managed in ways that were not very destructive to the environment. Out of numerous such healthy practices, one was three-field system wherein the peasants divided their farmland into three fields, one for winter crops, one for summer crops, and one to remain fallow. The use of the fields was rotated each year. A second part of the system, in order to prevent soil exhaustion, was to use different crops that took different nutrients from the soil. The winter crop typically would consist of winter wheat or rye, and the spring crop would be either spring wheat or legumes (beans or peas). The greater variety of crops provided people with a more balanced diet. Also an advantage of legumes is that they take nitrogen out of the air rather than the soil, and when buried, actually replenish the soil with nitrogen (the Romans referred to this as "green manuring"). Pastures, forests, and water resources were held in common, and their use was carefully regulated by village councils.

*Bhagavan : Objection is that the people have become so impatient for sense gratification, they have no patience anymore. They can't wait... There was some story. In the United States, there has been this trouble with petrol, and... All over the world, there's been this trouble with petrol, gasoline. So there was rationing. That means people could only get a little gas. So the cars would line up for a great distance in the gas station, and they'd wait for a long time. And sometimes the gas station would run out of gas. And the people would get so angry that they killed the gas station attendant...does not teach anyone to be austere or patient.*

*Prabhupada: But human life is meant for austerity and patience. Tapo divyam [SB 5.5.1]. Austerity, penance, that is human life. Otherwise, it is animal life. Simply animal civilization. (Conversation, June 12, 1974, Paris)*

The impact of the new methods of commercial agriculture on North European ecology was profound. Inhabitants came to perceive of their physical surroundings in basically capitalist terms. Natural resources increasingly were viewed as commodities, articles of value capable of being exchanged for other goods or money. Though ecological consequences varied according to region, every colony touched by the growing commercialisation suffered deforestation, epidemics, soil exhaustion, and decreasing numbers of wild animals. Market forces would continue to transform the European environment. Trees were cut to expand farmland and pasture and to supply fuel and raw materials for factories. Deforestation resulted in a drier landscape more vulnerable to erosion from high winds. Beaver, fox, and lynx had grown scarce as trappers and traders sought valuable pelts.

Medieval era began to see the sad demise of subsistence farming and introduction of cash crops. Inland communities with little access to markets practiced traditional agriculture that aimed to feed, clothe, and reproduce the family. This form of subsistence farming was far more ecologically sensitive than farming for the market would later be. After clearing forest trees by cutting or burning, farmers used small lots for crops for just a few years, rotating corn, beans, and squash between three fields. Those fields then lay fallow (unused) or served as pastureland for up to eight years, then reverted to forest while a new lot was cleared for the growing of crops. Such methods worked effectively to preserve soil nutrients as mentioned earlier.

Single-crop fields were more vulnerable to pests including insects, squirrels, and crows. Deforestation altered the climate resulting in colder springs, warmer summers, and earlier frosts. Planters, slaves, and small farmers all suffered from changes in the disease environment. As the aedes mosquito found breeding grounds in new ditches and reservoirs, populous towns endured epidemics of yellow fever and malaria.

*The fact is, we cannot drill our way to oil independence.*
*~Carl Pope*

Construction of roads and canals provided backcountry easier access to markets. The transportation and market revolutions altered the environment in two kinds of ways. Direct consequences included disruptions to the fragile ecosystems of rivers and lakes by canal and dam construction and the burning of vast quantities of firewood aboard new steamboats. Indirect consequences were perhaps more profound. New forms of transportation helped create new regions and economic zones.

## Oil Crisis- Gift of Un-Ecofriendly Science-Based Civilization

Like any monoculture (an agricultural system dominated by a single crop), single-crop fields promoted the development of soil toxins and the rapid multiplication of parasites. With access to better transportation, farmers began to participate in the market economy in new ways, beyond raising cash crops, that the landscape could not long sustain.

*Human life is never meant for sense gratification, but for self-realization. Srimad-Bhagavatam instructs us solely on this subject from the very beginning to the end. Human life is simply meant for self-realization. The civilization which aims at this utmost perfection never indulges in creating unwanted things, and such a perfect civilization prepares men only to accept the bare necessities of life or to follow the principle of the best use of a bad bargain. Our material bodies and our lives in that connection are bad bargains because the living entity is actually spirit, and spiritual advancement of the living entity is absolutely necessary. Human life is intended for the realization of this important factor, and one should act accordingly, accepting only the bare necessities of life and depending more on God's gift without diversion of human energy for any other purpose, such as being mad for material enjoyment. The materialistic advancement of civilization is called "the civilization of the demons," which ultimately ends in wars and scarcity.          -Srila Prabhupada (Srimad Bhagavatam 2.2.3)*

The eventual ecological decline of farms helped set the stage for early industrialization, which in turn created new environmental challenges. As farms faltered, many landless sons and daughters turned to wage labor in new factories including textile mills and sawmills. This new source of cheap labor, combined with the introduction of the power loom fueled an explosive textile industry. Sawmills also expanded, depleting forests. Construction of dams for the new industries altered the ecology of rivers in which fish, including salmon, were blocked from upstream spawning grounds. By the late 18th century, the signs of modern industrial pollution were already evident. As textile mills turned to steam power, burning coal, smoke blackened the skies over fast-growing cities.

The real onset of industrialization would have to await the railroad and textile boom of the Early nineteenth century.

During the seventeenth century, England, faced with a shortage of wood, had switched to coal as an energy source for industry. "Whereas the medieval economy had been based on organic and renewable energy sources-wood, water, and wind-the emerging capitalist economy taking shape over most of western Europe was based not only on the non-renewable energy source - coal - but on an inorganic economic core - metals: iron, copper, silver, gold, tin, and mercury - the refining and processing of which ultimately depended on and further depleted the forests," says Carolyn Merchant. Thus intensive agriculture began to destroy medieval field systems as well as the hedgerows and field trees.

The interaction between humankind and the environment was reciprocal: short-term effects of weather and longer-term climatic change, had profound consequences for medieval economies, societies, and cultures.

This un-ecofriendly science-based civilizational model that originated a few centuries ago in Europe would eventually spread all over the world in the form of 'American dream' and threaten the very existence of life on this planet. We can clearly see this happening now.

## From A Sensate Culture To Ideational Culture

Godless materialism of medieval and Renaissance Europe eventually gave rise to a civilization which was mad for maximum exploitation of matter. This paradigm owes the first and foremost responsibility for the oil crunch we face now. Therefore solution also lies in reducing the material fever of human society.

Unfortunately, this culture teaches an entirely different set of virtues. It emphasizes a self-centered, consumption-oriented lifestyle, which works directly against a nature friendly life style. It also places an unhealthy emphasis on living within the moment, rather than committing to long-term projects of personal discipline and learning.

In 1941, Harvard sociologist Pitirim Sorokin wrote a book entitled The Crisis of Our Age. In it Sorokin claimed that cultures come in two major types: sensate and ideational. A sensate culture is one in which people only believe in the reality of the physical world we experience with our five senses. A sensate culture is secular, this-worldly, and empirical. It believes that beyond the reality and values which we can see, hear, smell, touch, and taste there is no other reality and no real values. By contrast, an ideational culture embraces the physical world, but goes on to accept the notion that a non-physical, immaterial reality can be known as well, a reality consisting of God, the soul, immaterial beings, values and purposes. Sorokin claimed that a sensate culture will eventually disintegrate because it does not have the intellectual resources necessary to sustain a public and private life conducive to human flourishing. After all, if we can't know anything about values, life after death, God, and so forth, where can we find solid guidance toward a life of wisdom and character?

Proverbs tells us that we become the ideas we cherish in our inner

*...increasing the problem! They have to dig out petroleum oil from the midst of the ocean. Is it easy job?*
*Jayatirtha : No.*
*Prabhupāda: But they will do it because they have got motorcars, they must find out petrol.*
*(Morning Walk, September 28, 1972, Los Angeles)*

being and we transform our lives accordingly. Scripture is quite clear that our worldview will determine the shape of our cultural and individual lives. Because this is so, the worldview struggle raging in our modern context has absolutely far-reaching and crucial implications.

Sorokin further said that sensate society "intensely cultivates scientific knowledge of the physical and biological properties of sensory reality.....Despite its lip service to the values of the Kingdom of God, it cares mainly about the sensory values of wealth, health, bodily comfort, sensual pleasures, and lust for power and fame. Its dominant ethic is invariably utilitarian and hedonistic."

"The inevitable result, is the exceptional violence we have experienced in the twentieth century. And we may include in this category violence against the planet itself, brought on by the increasing destructiveness of the morally irresponsible, sensate scientific achievements ... invented and continuously perfected by the sensate scientists."

5000 years ago, Bhagavad gita defined this phenomenon as demonism, "They say that this world is unreal, with no foundation, no God in control. They say it is produced of sex desire and has no cause other than lust. Following such conclusions, the demoniac, who are lost to themselves and who have no intelligence, engage in unbeneficial, horrible works meant to destroy the world. (Bhagavad-gita 16.8-9)

## Focus On Interior Life

In the last few decades people have become far more concerned about external factors such as the possession of consumer goods, celebrity status, image, and power rather than the development of what is called an interior life. It wasn't long ago that people were measured by the internal traits of virtue and morality, and it was the person who exhibited character and acted honorably who was held in high esteem. This kind of life was built upon contemplation of what might be called the "good life." After long deliberation, an individual then disciplined himself in those virtues most valued.

Paul reminds us of the dangers of over-emphasizing personality ethics as opposed to character ethics when he writes, "Their destiny is destruction, their god is their stomach, and their glory is in their shame. Their mind is on earthly things.

The oil crisis is there because we're conditioned to see the earth and its resources and creatures as things to be exploited unlimitedly for personal gratification. And this is so because we have forgotten our connection with the transcendental reality.

## Attitude Towards Nature

There is a growing understanding that addressing the global crisis facing humanity will require new methods for knowing, understanding and valuing the world. Narrow, disciplinary, mechanistic, and reductionist perceptions of reality are proving inadequate for addressing the complex, interconnected problems of the current age. The currently dominant worldview of scientific materialism, which views the cosmos as a vast machine composed of independent, externally related pieces, promotes fragmentation in our thinking and perception. The materialist view of natural systems as commodities to be exploited coupled with the ethos of consumerism and social Darwinism has encouraged widespread destruction of our natural life support systems. The cancerous spread of nihilism and dehumanization are driving the decay and disintegration of techno-industrial culture.

The paradigm of reductionism gave human society increased ability to exploit matter. But this viewpoint has cost us a lot - a planet threatened with ruination, and depletion of the human spirit.

In Vedic literatures like Bhagavad-gita, material nature has been described as God's inferior energy and living entity, the soul as God's superior nature. Vedic literatures do not approve the idea that life can originate from chemicals and neither does empirical science has any proof to support this idea. By portraying life forms to be mere biological machines, science has done tremendous damage. This godless perspective lies at the heart of all the maladies including environmental ones. True God consciousness inspires one to treat environment as one of the God's manifestations and act responsibly towards it. Godless worldview has produced an extremely callous attitude towards ecology and world resources.

Ideas or thoughts do manifest as our destiny as reflected in the adage, "Sow a thought to reap an act, sow an act and reap a habit, sow a habit and reap a character, sow a character and reap a destiny."

# Solution -3

## Deep, Spiritual Change Of Heart

### Purpose And Destination
### Of Life

*A certain degree of physical comfort is necessary but above a certain level it becomes a hindrance instead of a help; therefore the ideal of creating an unlimited number of wants and satisfying them, seems to be a delusion and a trap. The satisfaction of one's physical needs must come at a certain point to a dead stop before it degenerates into physical decadence. Europeans will have to remodel their outlook if they are not to perish under the weight of the comforts to which they are becoming slaves.*

*~Mahatma Gandhi*

## Living Gently
## From Blind Materialism To Spiritual Development

Rampant blind materialism lies at the root of petroleum crisis. We are recklessly using up a precious source of energy in a mere 150 years which took nature millions of years to create. Such rate of consumption, to put it mildly, is nothing short of mass insanity. The solution lies at turning our face towards spiritual development, once again.

Its heartening to see that several influential environmental organizations in the world, though secular by constitution, are echoing this theme.

Once such forum is the Worldwatch Institute, an independent research organization that works for an environmentally sustainable and socially just society, in which the needs of all people are met without threatening the health of the natural environment or the well-being of future generations.

Alan Durning of Worldwatch Institute writes, "In a fragile biosphere, the ultimate fate of humanity may depend on whether we can cultivate a deeper sense of self-restraint, founded on a widespread ethic of limiting consumption and finding non-material enrichment...Those who seek to rise to this environmental challenge may find encouragement in the body of human wisdom passed down from antiquity. To seek out sufficiency is to follow the path of voluntary simplicity preached by all the sages from Buddha to Mohammed. Typical of these pronouncements is this passage from the Bible: "What shall it profit a man if he shall gain the whole world and lose his own soul?"

Allen adds,"....action is needed to restrain the excesses of advertising, to curb the shopping culture, and to revitalize household and community economies as human-scale alternatives to the high consumption lifestyle. There could be many more people ready to begin saying "enough"....After all, much of what we consume is wasted or unwanted in the first place. How much of the packaging that wraps products we consume each year - 162 pounds per capita in the United States - would we rather not see? ...How many of the unsolicited sales pitches each American receives each day in the mail - 37 percent of all mail - are nothing but bothersome junk? How much of the advertising in our morning newspaper - covering 65 percent of the newsprint in American paper - would we not gladly see left out?"

Allen continues, "How many of the miles we drive - almost 6,000 a year a piece in the United States - would we not happily give up if livable neighborhoods were closer to work, a variety of local merchants closer to home, streets safe to walk and bicycle, and public transit easier and faster?

Keith C. Heidorn sums this up when he defines 'Living Gently' as a voluntary manner of living which pursues a positive, satisfying life that is considerate, noble and easily managed and that seeks to produce as small an impact on the environment as possible. It is a lifestyle chosen not only for personal satisfaction, but also for the good of our fellow inhabitants of Planet Earth: animals, humans and plants. It involves frugality but goes beyond.

## Overall Spiritualization of Society

### Bad Theology Creates Bad Ecology

Humanity is on a "spiral to suicide" and that the environmental discourses of academia often suggest an 'end-of-the-pipeline approach'. Mary Evelyn Tucker echoes this misgiving. "We're all concerned about simply rhetorical statements or a superficial approach [that] is not going to tap into the deep spiritual reservoirs of people.

*Everyone is engaged in producing motor tire, car, and they are flattering the Arabians for petrol. The same energy, if it would have been engaged in producing food grain, then where is the poverty? -Srila Prabhupada (Room Conversation with Press reporters, March 21, 1975, Calcutta)*

A spiritual change of heart could only offer a solution towards this 'much talked about and little done for' crisis.

## Ideas Have Consequences

Adolf Hitler had ideas that he expressed in a book entitled "Mein Kampf." Karl Marx had ideas that he expressed in "Das Kapital." Both sets of ideas have had enormous consequences for human history.

Our environmental crisis has its roots in incorrect and imperfect idea of the self and the universe. When we understand our true spiritual nature, our unlimited urge to consume things and to produce things for consumption can be curbed. The natural result will be a better environment in which to pursue spiritual growth instead of excessive economic growth.

The theologian Jürgen Moltmann wrote that the "alienation of nature brought about by human beings can never be overcome until men find a new understanding of themselves and a new interpretation of their world in the framework of nature."

With a deep, spiritual change of heart, a permanent change of goals and values, environmental reform would take place as a by-product, almost automatically.

## Philosophical Dimensions Of The Problem.

One difficulty is that most individual and collective attempts fail to recognize the philosophical dimensions of the problem.

Says Vanamali Dasa, former head of the 225-acre Hare Krishna community near the town of Czarnow, Poland. "The biggest problem in society today is that almost all of us claim God's property as our own. By trying to possess and enjoy more than our rightful share of this planet, we continuously act against higher laws of the universe. One result is unclean industrial enterprises and factories that create enormous amounts of waste, pollution, noise, and anxiety."

Another tragic irony is that, while modern society has been highly effective in producing material goods, it has failed to provide us with a deeper sense of fulfillment. Consumer society's excessive use of throwaway food and beverage containers, for example, is as much an economic, cultural and spiritual issue as an environmental one. We can no longer afford to isolate our problems and our solutions, they are all interrelated. As the late futurist Willis Harman suggested, we need to address the "systemic failures" of industrial civilization head on. To heal consumer society's wounds - including its environmental damages, cultural decay, economic disparity, and spiritual shallowness - we must examine and treat it as a complete organism, much the same way holistic medicine attempts to restore the whole individual.

But environmental, political, economic or cultural changes are not enough. A truly holistic vision for both people and planet must include Cosmic Consciousness or God - the source from which everything originates, maintainer of everything be and in whom everything rests.

## Pollution Of Consciousness And Pollution Of The Environment

Humanity has reached a turning point, a defining moment in history. We stand at a crossroads, and the path we choose to follow will affect future generations at least as profoundly as the industrial revolution affected our lives. The main problems of modern planetary civilization can no longer be solved in isolation. Environmental destruction, cultural decay, and technological excesses, along with increased poverty, even in the world's richest nations - are all systemic problems that cannot be changed with the fragmented approaches so far employed by politicians and scientists. We need bold, new, comprehensive concepts and visions.

But most of all, we need a change of heart. As former president of Czechoslovakia Vaclav Havel said before a Joint Session of the United States Congress in 1990: "Without a global revolution in the sphere of human consciousness a more human society will not emerge."

## Science Joins The Holy Bandwagon

Moscow hosted, in January 1990, an environmental meet called **Global Forum of Spiritual and Parliamentary Leaders.** Practically for the first time, some of the world's leading scientists endorsed need for a spiritual solution to our planet's resource and environmental crisis. Thirty-two leading scientists signed a document entitled "Preserving and Cherishing the Earth: An Appeal for Joint Commitment in Science and Religion."

The signers included astronomer Carl Sagan, nuclear winter theorist Paul J. Crutzen, physicist Freeman J. Dyson, paleontologist Stephen J. Gould, environmental scientist Roger Revelle, and former Massachusetts Institute of Technology (MIT) president Jerome Wiesner.

The scientists said, "We are close to committing - many would argue we are already committing - what in religious language is sometimes called Crimes against Creation." They therefore issued an urgent "appeal to the world religious community to commit, in word and deed, and as boldly as is required, to preserve the environment of the earth."

"The environmental crisis requires radical changes not only in public policy," said the scientists, "but also in individual behavior. The historical record makes clear that religious teaching, example, and leadership are powerfully able to influence personal conduct and commitment."

"As scientists, many of us have had profound experiences of awe and reverence before the universe," added the signers. "We understand that what is regarded as sacred is more likely to be treated with care and respect. Our planetary home should be so regarded. Efforts to safeguard and cherish the environment need to be infused with a vision of the sacred."

On November 18th, 1992, some five months after the largest gathering of heads of state in history at the Earth Summit in Rio, a document was released entitled "World Scientists' Warning to Humanity." It was signed by more than sixteen hundred senior scientists from seventy-one countries including half of all Nobel Prize winners still alive at the time. It was one of the most unprecedented consensus ever made by the intellectual elite in the world. The warning was eloquent but blunt. No more than a few decades remain before current human practices make human life, as we know it on the planet today, unsustainable.

# Solution -4

## Voluntary Simplicity

### (LOVOS - Lifestyles of Voluntary Simplicity)

*It is by ignorance that people think that by opening factories they will be happy. Why should they open factories? There is no need. There is so much land, and one can produce one's own food grains and eat sumptuously without any factory. Milk is also available without a factory. The factory cannot produce milk or grains. When everyone is working in the city to produce nuts and bolts, who will produce food grains? Simple living and high thinking is the solution to economic problems. Therefore the Krsna consciousness movement in engaging devotees in producing their own food and living self-sufficiently so that rascals may see how one can live very peacefully, eat the food grains one has grown oneself, drink milk, and chant Hare Krsna.*
*– Srila Prabhupada (TQK 18: Liberation from Ignorance and Suffering)*

## Voluntary Simplicity & Non-Material Satisfaction

In the U.S. there are 1,000 cars for every 1,000 adults.
In Germany, it's 550 cars for every 1,000 adults.
In India, there are four cars for every 1,000 adults.

Let us take a look at other figures. The developed world consumes about 4 tonnes of crude oil per year per capita. Within the developed world, North Americans annually consume about 340 Gigajoules per person (GJ/person), whereas Europeans consume 150 GJ/person of commercial energy. Most of the rest of the world consumes about 60 GJ/person, and India and China consume 10 and 30 GJ/person, respectively.

So technically it is possible to live with less resources and lesser damage to the environment. When one voluntarily decides to consume less, it is called voluntary simplicity. Voluntary simplicity means doing/having/living more with less - more time, meaning, joy, satisfaction, relationships, community; less money, material possessions, stress, competition, isolation. Voluntary simplicity is a growing movement of people who have realized that happiness and fulfillment do not lie in having more money, or new and bigger things.

So in this type of lifestyle individuals consciously choose to minimize the 'more-is-better' pursuit of wealth and consumption. Adherents choose simple living for a variety of reasons, including spirituality, health, increase in 'quality time' for family and friends, stress reduction, conservation, social justice or anti-consumerism, while others choose to live more simply for reasons of personal taste or personal economy.

Simple living as a concept is distinguished from those living in forced poverty, as it is a voluntary lifestyle choice. Although asceticism resembles voluntary simplicity, proponents of simple living are not all ascetics. The term "downshifting" is often used to describe the act of moving from a lifestyle of greater consumption towards a lifestyle based on voluntary simplicity.

In sharp contrast with consumeristic American way of life, here we can quote few passages from Srimad Bhagavatam, an ancient vedic text, to show how in bygone ages they viewed voluntary simplicity.

SB 2.2.3

atah kavir namasu yavad arthah
syad apramatto vyavasaya-buddhih
siddhe 'nyatharthe na yateta tatra
parisramam tatra samiksamanah

For this reason the enlightened person should endeavor only for the minimum necessities of life while in the world of names. He should be intelligently fixed and never endeavor for unwanted things, being competent to perceive practically that all such endeavors are merely hard labor for nothing.

PURPORT

The bhagavata-dharma, or the cult of Srimad-Bhagavatam, is perfectly distinct from the way of fruitive activities, which are considered by the devotees to be merely a waste of time. The whole universe, or for that matter all material existence, is moving on as jagat, simply for planning business to make one's position very comfortable or secure, although everyone sees that this existence is neither comfortable nor secure and can never become comfortable or secure at any stage of development. Those who are captivated by the illusory advancement of material civilization (following the way of phantasmagoria) are certainly madmen. The whole material creation is a jugglery of names only; in fact, it is nothing but a bewildering creation of matter like earth, water and fire. The buildings, furniture, cars, bungalows, mills, factories, industries, peace, war or even the highest perfection of material science, namely atomic energy and electronics, are all simply bewildering names of material elements with their concomitant reactions of the three modes. Since the devotee of the Lord knows them perfectly well, he is not interested in creating unwanted things for a situation which is not at all reality, but simply names of no more significance than the babble of sea waves. The great kings, leaders and soldiers fight with one another in order to perpetuate their names in history. They are forgotten in due course of time, and they make a place for another era in history. But the devotee realizes how much history and historical persons are useless products of flickering time. The fruitive worker aspires after a big fortune in the matter of wealth, woman and worldly adoration, but those who are fixed in perfect reality are not at all interested in such false things. For them it is all a waste of time. Since every second of human life is important, an enlightened man should be very careful to utilize time very cautiously. One second of human life wasted in the vain research of planning for happiness in the material world can never be replaced, even if one spends millions of coins of gold. Therefore, the transcendentalist desiring freedom from the clutches of maya, or the illusory activities of life, is warned herewith not to be captivated by the external features of fruitive actors. Human life is never meant for sense gratification, but for self-realization. Srimad-Bhagavatam instructs us solely on this subject from the very beginning to the end. Human life is simply meant for self-realization. The civilization which aims at this utmost perfection never indulges in creating unwanted things, and such a perfect civilization prepares men only to accept the bare necessities of life or to follow the principle of the best use of a bad bargain. Our material bodies and our lives in that connection are bad bargains because the living entity is actually spirit, and spiritual advancement of the living entity is absolutely necessary. Human life is intended for the realization of this important factor, and one should act accordingly, accepting only the bare necessities of life and depending more on God's gift without diversion of human energy for any other purpose, such as being mad for material enjoyment. The materialistic advancement of civilization is called "the civilization of the demons," which ultimately ends in wars and scarcity. The transcendentalist is specifically warned herewith to be fixed in mind, so that even if there is difficulty in plain living and high thinking he will not budge even an inch from his stark determination. For a transcendentalist, it is a suicidal policy to be intimately in touch with the sense-gratifiers of the world, because such a policy will frustrate the ultimate gain of life. Sukadeva Gosvami met Maharaja Pariksit when the latter felt a necessity for such a meeting. It is the duty of a transcendentalist to help persons who desire real salvation and to support the cause of salvation. One might note that Sukadeva Gosvami never met Maharaja Pariksit while he was ruling as a great king. For a transcendentalist, the mode of activities is explained in the next sloka.

SB 2.2.4

satyam ksitau kim kasipoh prayasair
bahau svasiddhe hy upabarhanaih kim
saty anjalau kim purudhanna-patrya

dig-valkaladau sati kim dukulaih

When there are ample earthly flats to lie on, what is the necessity of cots and beds? When one can use his own arms, what is the necessity of a pillow? When one can use the palms of his hands, what is the necessity of varieties of utensils? When there is ample covering, or the skins of trees, what is the necessity of clothing?

PURPORT

The necessities of life for the protection and comfort of the body must not be unnecessarily increased. Human energy is spoiled in a vain search after such illusory happiness. If one is able to lie down on the floor, then why should one endeavor to get a good bedstead or soft cushion to lie on? If one can rest without any pillow and make use of the soft arms endowed by nature, there is no necessity of searching after a pillow. If we make a study of the general life of the animals, we can see that they have no intelligence for building big houses, furniture, and other household paraphernalia, and yet they maintain a healthy life by lying down on the open land. They do not know how to cook or prepare foodstuff, yet they still live healthy lives more easily than the human being. This does not mean that human civilization should revert to animal life or that the human being should live naked in the jungles without any culture, education and sense of morality. An intelligent human cannot live the life of an animal; rather, man should try to utilize his intelligence in arts and science, poetry and philosophy. In such a way he can further the progressive march of human civilization. But here the idea given by Srila Sukadeva Gosvami is that the reserve energy of human life, which is far superior to that of animals, should simply he utilized for self-realization. Advancement of human civilization must be towards the goal of establishing our lost relationship with God, which is not possible in any form of life other than the human. One must realize the nullity of the material phenomenon, considering it a passing phantasmagoria, and must endeavor to make a solution to the miseries of life. Self-complacence with a polished type of animal civilization geared to sense gratification is delusion, and such a "civilization" is not worthy of the name. In pursuit of such false activities, a human being is in the clutches of maya, or illusion. Great sages and saints in the days of yore were not living in palatial buildings furnished with good furniture and so-called amenities of life. They used to live in huts and groves and sit on the flat ground, and yet they have left immense treasures of high knowledge with all perfection. Srila Rupa Gosvami and Srila Sanatana Gosvami were high-ranking ministers of state, but they were able to leave behind them immense writings on transcendental knowledge, while residing only for one night underneath one tree. They did not live even two nights under the same tree, and what to speak of well-furnished rooms with modern amenities. And still they were able to give us most important literatures of self-realization. So-called comforts of life are not actually helpful for progressive civilization; rather, they are detrimental to such progressive life. In the system of sanatana-dharma, of four divisions of social life and four orders of progressive realization, there are ample opportunities and sufficient directions for a happy termination of the progressive life, and the sincere followers are advised therein to accept a voluntary life of renunciation in order to achieve the desired goal of life. If one is not accustomed to abiding by the life of renunciation and self-abnegation from the beginning, one should try to get into the habit at a later stage of life as recommended by Srila Sukadeva Gosvami, and that will help one to achieve the desired success.

SB 2.2.5

cirani kim pathi na santi disanti bhiksam
naivanghripah para-bhrtah sarito 'py asusyan
ruddha guhah kim ajito 'vati nopasannan
kasmad bhajanti kavayo dhana-durmadandhan

Are there no torn clothes lying on the common road? Do the trees, which exist for maintaining others, no longer give alms in charity? Do the rivers, being dried up, no longer supply water to the thirsty? Are the caves of the mountains now closed, or, above all, does the Almighty Lord not protect the fully surrendered souls? Why then do the learned sages go to flatter those who are intoxicated by hard-earned wealth?

PURPORT

The renounced order of life is never meant for begging or living at the cost of others as a parasite. According to the dictionary, a parasite is a sycophant who lives at the cost of society without making any contribution to that society. The renounced order is meant for contributing something substantial to society and not depending on the earnings of the householders. On the contrary, acceptance of alms from the householders by the bona fide mendicant is an opportunity afforded by the saint for the tangible benefit of the donor. In the sanatana-dharma institution, alms-giving to the mendicant is part of a householder's duty, and it is advised in the scriptures that the householders should treat the mendicants as their family children and should provide them with food, clothing, etc., without being asked.

Pseudomendicants, therefore, should not take advantage of the charitable disposition of the faithful householders. The first duty of a person in the renounced order of life is to contribute some literary work for the benefit of the human being in order to give him realized direction toward self-realization. Amongst the other duties in the renounced order of life of Srila Sanatana, Srila Rupa and the other Gosvamis of Vrndavana, the foremost duty discharged by them was to hold learned discourses amongst themselves at Sevakunja, Vrndavana (the spot where Sri Radha-Damodara Temple was established by Srila Jiva Gosvami and where the actual samadhi tombs of Srila Rupa Gosvami and Srila Jiva Gosvami are laid). For the benefit of all in human society, they left behind them immense literatures of transcendental importance. Similarly, all the acaryas who voluntarily accepted the renounced order of life aimed at benefiting human society and not at living a comfortable or irresponsible life at the cost of others. However, those who cannot give any contribution should not go to the householders for food, for such mendicants asking bread from the householders are an insult to the highest order. Sukadeva Gosvami gave this warning especially for those mendicants who adopt this line of profession to solve their economic problems. Such mendicants are in abundance in the age of Kali. When a man becomes a mendicant willfully or by circumstances, he must be of firm faith and conviction that the Supreme Lord is the maintainer of all living beings everywhere in the universe. Why, then, would He neglect the maintenance of a surrendered soul who is cent percent engaged in the service of the Lord? A common master looks to the necessities of his servant, so how much more would the all-powerful, all-opulent Supreme Lord look after the necessities of life for a fully surrendered soul. The general rule is that a mendicant devotee will accept a simple small loincloth without asking anyone to give it in charity. He simply salvages it from the rejected torn cloth thrown in the street. When he is hungry he may go to a magnanimous tree which drops fruits, and when he is thirsty he may drink water from the flowing river. He does not require to live in a comfortable house, but should find a cave in the hills and not be afraid of jungle animals, keeping faith in God, who lives in everyone's heart. The Lord may dictate to tigers and other jungle animals not to disturb His devotee. Haridasa Thakura, a great devotee of Lord Sri Caitanya, used to live in such a cave, and by chance a great venomous snake was a co-partner of the cave. Some admirer of Thakura Haridasa who had to visit the Thakura every day feared the snake and suggested that the

Thakura leave that place. Because his devotees were afraid of the snake and they were regularly visiting the cave, Thakura Haridasa agreed to the proposal on their account. But as soon as this was settled, the snake actually crawled out of its hole in the cave and left the cave for good before everyone present. By the dictation of the Lord, who lived also within the heart of the snake, the snake gave preference to Haridasa and decided to leave the place and not disturb him. So this is a tangible example of how the Lord gives protection to a bona fide devotee like Thakura Haridasa. According to the regulations of the sanatana-dharma institution, one is trained from the beginning to depend fully on the protection of the Lord in all circumstances. The path of renunciation is recommended for acceptance by one who is fully accomplished and fully purified in his existence. This stage is described also in the Bhagavad-gita (16.5) as daivi sampat. A human being is required to accumulate daivi sampat, or spiritual assets; otherwise, the next alternative, asuri sampat, or material assets, will overcome him disproportionately, and thus one will be forced into the entanglement of different miseries of the material world. A sannyasi should always live alone, without company, and he must be fearless. He should never be afraid of living alone, although he is never alone. The Lord is residing in everyone's heart, and unless one is purified by the prescribed process, one will feel that he is alone. But a man in the renounced order of life must be purified by the process; thus he will feel the presence of the Lord everywhere and will have nothing to fear (such as being without any company). Everyone can become a fearless and honest person if his very existence is purified by discharging the prescribed duty for each and every order of life. One can become fixed in one's prescribed duty by faithful aural reception of Vedic instructions and assimilation of the essence of Vedic knowledge by devotional service to the Lord.

The act of voluntary simplicity or cultivation of detachment has been the focal point of all religious teachings. In human history, there have been countless souls who led a life of selflessness, dedication and voluntary simplicity. These personalities are respected, revered and remembered even today.

Also known as downshifting, many today are deciding to reduce their incomes and place family, friends and contentment above money in determining their life goals.

# Solution -5

## Science Of Consciousness
## Bhagavad-gita

*The basic principle of this modern civilization is wrong. Everyone, the so-called advanced scientists, so-called advanced philosopher or politician, everyone is thinking that "I am this body." So on the basic principle they're wrong. Therefore the so-called advancement of civilization is wrong. It's... At one point mathematical calculation, if you have done mistake in one point... Two plus two equal two. Why if you have made up three, the mistake, then the whole calculation will be mistaken. The balance, it will never tally. Similarly, our present civilization... Not present; it is always there. Now it is very strong bodily conception of life, so the basic principle is wrong. Therefore whatever we are advancing, that is wrong. Parabhava, defeat. That is stated in the Srimad-Bhagavatam. Basic principle is wrong, abodha, ignorance.*
*- Srila Prabhupada*
*(Lecture, Srimad-Bhagavatam - Los Angeles, December 9, 1973)*

## Bhagavad-gita - Voice Of An Old Intelligence
### Fundamental Question: Who am I?

Environmental, political, economic or cultural changes and upheavals are not enough. A truly holistic vision for both people and planet must include clear idea as to one's real identity.

The voices of many dissenting experts echo the deep sentiments of millions of people all over the world. Most Americans, for example, identify themselves as environmentalists. More and more households recycle paper, glass and soda containers, buy energy-efficient light bulbs, and donate money to help save the whales and the rainforest. On Capitol Hill, in the courts, and in the streets, the environmental movement has enjoyed many victories. Yet, despite all these accomplishments, the environment is in serious trouble, and the problems are getting worse with each passing year. For this increasingly worsening problem, here lies the solution: Science of consciousness.

"As the sun alone illuminates all this universe, so does the living entity, one within the body, illuminate the entire body by consciousness." *(Bhagavad-gita 13.34)*

If the world is ever to become free from the threat of environmental annihilation, we shall have to undertake a thorough reexamination of the materialistic assumptions underlying not only our picture of nature but our conception of our very selves.

Some scientists are already beginning to question whether materialistic principles are really adequate to explain basic features of human existence-such as consciousness. For example, John C. Eccles, a Nobel-prize-winning neurobiologist, states, "The ultimate problem relates to the origin of the self, how each of us as a self-conscious being comes to exist as a unique self associated with a brain. This is the mystery of personal existence." Eccles said that "the uniqueness each of us experiences can be sufficiently explained only by recourse to some supernatural origin."

Thomas Berry mentions that we must reestablish a spiritual "intimacy" with the earth as compared to our overly scientific-technological relationship with the earth. That is only possible when we understand that we are essentially spirit souls and not bodies.

## Consumption vs. Self-realization

If the conscious self is factually supernatural in origin, and if this knowledge were firmly integrated into our educational and cultural institutions, society would probably be much more directed toward self-realization than it is today. The overwhelming impetus toward the domination and exploitation of matter that underlies today's industrial civilization and culminates in resource crunch would certainly be lessened.

In words of Grandon Harris: Our honeymoon with the planet earth is over. We must take our marriage with the earth seriously. We cannot divorce it, but it can divorce us!

This seriousness, as opposed to frivolousness, comes from understanding our real identity as spirit souls and not as Darwinian monkeys. From a monkey, hardly any seriousness can be expected.

Following verses from Bhagavad-gita illustrate these points.

kamopabhoga-parama, etavad iti niscitah

They believe that to gratify the senses is the prime necessity of human civilization. (Bg 16.11) This verse of Bhagavad-gita tells us that those who mistake their identities or those who fail to understand who they are, they tread the path of sense aggrandizement. Petroleum crisis is an outcome of this viewpoint.

Another verse (Bg 2.62) sums up the our consumeristic civilization:

dhyayato visayan pumsah

sangas tesupajayate

sangat sanjayate kamah

kamat krodho 'bhijayate

While contemplating the objects of the senses, a person develops attachment for them, and from such attachment lust develops, and from lust anger arises.

This verse of Bhagavad-gita tells us that by contemplating the objects of the senses one becomes attached to them and ultimately ends up frustrated and bewildered. Industrialized society in particular has as its cornerstone the need to stimulate consumption, to constantly fuel economic growth. To this end, it constantly encourages us to meditate on the objects of our senses. With individuals' desires massively outstripping their abilities to meet their aspirations, it is hardly any surprise that we create ongoing frustration and extreme egotism, which result in environmental, social, and cultural devastation.

Text 4.22, 'yadrccha-labha-santusto' describes that by practicing bhakti-yoga, one is satisfied with gain which comes of its own accord and one attains a taste for simple living and high thinking. In other words, developing love for God automatically moderates one's appetite for material things by enriching one's life spiritually.

## Science Of Consciousness

Great thinkers, both in the past and present have echoed the teachings of Bhagavad-gita. For example Plato believed in the immortality of the human soul. The soul was, he thought, an entity that was fundamentally distinct from the body although it could be and often was affected by its association with the body, being dragged down by what he called in one passage "the leaden weights of becoming." The soul was simple, not composite, and thus not liable to dissolution as were material things; further, it had the power of self-movement, again in contrast to material things. Ideally the soul should rule and guide the body, and it could ensure that this situation persisted by seeing that the bodily appetites were indulged to the minimum extent necessary for the continuance of life. The true philosopher, as Plato put it in the Phaedo, made his life a practice for death because he knew that after death the soul would be free of bodily ties and would return to its native element.

Many eminent men of science have stated that life is not reducible

*These rascal scientists have no common sense. Where is the machine that is working without any operator? Is there such a machine within their experience? How can they suggest that nature is working automatically? Nature is a wonderful machine but the operator is God, Krishna. That is real knowledge. (Conversation- March 1975)*

to chemistry and physics. These include Alfred Wallace (co-author of Charles Darwin's first publication on evolution); Thomas H. Huxley (a contemporary of Darwin's who championed Darwin's evolutionary theory); and Nobel physicists Niels Bohr and Eugene Wigner. The eminent mathematician John von Neumann has shown how quantum mechanics implies that the consciousness of the observer (he called it the "abstract ego") is distinct from all aspects of the observer's body and brain. This concept of an "abstract ego" corresponds to the irreducible nonmaterial entity posited by the theory of production and called the jivatma by Lord Krishna in Bhagavad-gita.

The Gita (2.20, 2.17) offers extensive information about the nature of the nonmaterial particle that imparts the symptoms of life to the material body: "For the soul there is neither birth nor death at any time. He has not come into being, does not come into being, and will not come into being. He is unborn, eternal, ever-existing, and primeval. He is not slain when the body is slain.... That which pervades the entire body you should know to be indestructible. No one is able to destroy that imperishable soul."

But today, influenced by materialistic science's refusal to consider the existence of a nonmaterial conscious self, people tend to identify exclusively with the body and mind. They therefore tend to exploit matter for the purpose of continually increasing their bodily satisfaction. Expressed through today's urban-industrial civilization, this exploitation is causing environmental decay of unprecedented global proportions.

Glitter of industrialization has covered us, ie., our souls from our vision. Rediscovering our identity will make our God-centered relationship with the environment possible.

Understanding the difference between our temporary material identity and our true spiritual identity is the key to solving the

*"Any group of beings (human or nonhuman, plant or animal) who take more from their surroundings than they give back will, obviously, deplete their surroundings, after which they will either have to mend ways or perish."*

environmental crisis. The foundation for an environmentally healthy planet is a science of consciousness that incorporates knowledge of the soul.

Over the last few years, research into consciousness has at last become accepted within the academic community. As John Searle puts it, raising the subject of consciousness in cognitive science discussions is no longer considered to be ``bad taste", causing graduate students to ``roll their eyes at the ceiling and assume expressions of mild disgust."

The Gita offers a simple solution to environmental anomalies, linking our problems directly to our lack of spiritual culture and values. Forgetfulness of our spiritual nature is making us overuse the technology to meet the exaggerated demands of the senses, leading to resource crunch like oil squeeze.

# Solution -6

## Karma & Crisis of Shortage

*"Quite apart from the laws of physics and chemistry, as laid down in quantum theory, we must consider laws of quite a different kind."*
*–Niels Bohr, Nobel laureate In Physics*

Put in simple words, Law of Karma is all about reaping what we sow." Life evens out. We have to pay for our misdeeds.

Law of Karma has had quite a karma. Long after India's seers immortalized it in the Vedas, it suffered bad press under European missionaries who belittled it as "fate" and "fatalism." Today it finds itself again in the ascendancy as the subtle and all-encompassing principle which governs man's experiential universe in a way likened to gravity's governance over the physical plane. Like gravity, karma was always there in its fullest potency, even when people did not comprehend it.

Each of us as individuals, as well as each group and nation, is continually creating karma, both good and bad. All thoughts, words, or actions will sooner or later come full circle and return home to their creators. This is the universal Law of Cause and Effect, which is operating all the time, just like gravity, whether we choose to believe it or not.

"What goes around, comes around" is a statement of fact. An African tribe puts this in a colorful way: "He who excretes on the road, meets flies on his return." In the East, it is said that the wheels of karma grind slowly, but exceedingly fine.

The United States may have experienced the karmic consequences of its arming and training religious zealots to defeat the communists

*The stringent laws of nature, under the order of the Supreme Personality of Godhead, cannot be altered by any living entity. The living entities are eternally under the subjugation of the almighty Lord. The Lord makes all the laws and orders, and these laws and orders are generally called dharma or religion.*
*(Srimad Bhagavatam 1.8.4)*

in Afghanistan in the 1980s, and then abandoning support for them, as several of those it trained were reportedly involved in the World Trade Center bombing. Also contributing to the karma of terrorist attacks might be U.S. policies that support dictatorships, ignore human rights and contribute to the poverty and suffering of people.

So in the same way, the sufferings caused by shortage or higher prices of gas has something to do with our individual and collective karma.

Then solution is to avoid bad karma and engage in good karma.

Many see the resources and environmental problem as strictly technical, with technical solutions. Even those who see an ethical dimension may see solutions only in terms of mobilizing public opinion to change certain obvious environmentally destructive behaviors. But there are deeper dimensions to the environmental crisis as already examined in earlier chapters. The environmental crisis," one ecologist has noted, "is an outward manifestation of a crisis of mind and spirit."

In addition to the laws presently known to science, there are, according to the Vedic literature, higher-order laws that govern the interactions of conscious entities. These laws are collectively known as karma. In the Vedic literature, karma is described in terms of actions and reactions. For example, if one causes unnecessary suffering to

*Prammattah. Pra means prakrsta-rupena. Mattah, mad. We are living entities. We have come here in this material world for sense enjoyment, and we are therefore mad after it, prammattah. So nunam prammattah kurute vikarma. Vikarma means which is against the laws. Just like karma, akarma, vikarma. These are explained. So vikarma means against the law. The Vedic version, they give us that "You should work in this way." But if we do not act according to the Vedic injunctions, that is called vikarma. And we become subjected to sufferings, impious activities. But we do it because we are mad after sense gratification. We do not care. Just like a thief, he knows that by stealing he'll be punished, but still, because he's mad after stealing, he'll do it, taking the risk of being arrested and being harassed. ~Srila Prabhupada (Lecture, Nectar of Devotion - Vrndavana, November 4, 1972)*

another living entity, one will undergo suffering in return. This suffering may also come as a resource crunch.

Niels Bohr, also a Nobel laureate, stated, "All of us know that there is such a thing as consciousness, simply because we have it ourselves. Hence consciousness must be a part of nature, or, more generally, of reality, which means that quite apart from the laws of physics and chemistry, as laid down in quantum theory, we must consider laws of quite a different kind."

Science today exists not only in the ignorance of the unity of the various material fields it has invented, but also exists totally confused to comprehend the fields that exist beyond the matter, at the mind and consciousness level.

Along the same lines, physicist David Bohm says, "The possibility is always open that there may exist an unlimited variety of additional properties, qualities, entities, systems, levels, etc., to which apply correspondingly new kinds of laws of nature."

The feeling that consciousness is an integral part of the nature of this universe is becoming increasingly strong with advancement in quantum nature of reality. It is now strongly felt by the scientific community that the primary realities of the unified field and conscious field are inseparable aspects of the same underlying process and united through mutual participation. Since the unified field [material energy] permeates all manifest phenomenon, so should its complementary aspect, consciousness.

Karma plays a leading role in the world's drift toward environmental catastrophe, and a large part of this karma is generated by unnecessary killing.

The strict law of karma deals measure for measure with anyone who violates the laws of nature. Destroying nature, natural resources and life is definitely a bad karma. Humanity is committing so many heinous crimes against creation and it can not expect to go scot-free.

## Population & Karma

### Problem of Overconsumption, Not Overpopulation

In traditional cultures, the earth is considered a personality and is worshiped. In Vedic literature, earth is known as mother Bhumi and out of affection for her children, the world population, she produces all their requirements. This is definitely an eco-friendly idea.

According to India's Vedic teachings, the earth can always produce enough of life's necessities and scarcity is not caused by overpopulation but by the negative karma generated by self-destructive actions of the planet's population. When the population is engaged impiously, mother Bhumi (the earth) restricts supply.

In Greek culture, goddess of the earth is called Gaia. Nasa Scientist Jim Lovelock proposes that the earth's atmosphere was an extraordinary and unstable mixture of gases, yet it was constant in composition over quite long periods of time. Could it be that life on Earth not only made the atmosphere, but also regulated it - keeping it at a constant composition, and at a level favourable for organisms". He calls his hypothesis the Gaia principle.

Lovelock says, "The concept of Mother Earth, or, as the Greeks called her long ago, Gaia, has been widely held throughout history and has been the basis of a belief which still coexists with the great religions."

"Therefore," states Srila Prabhupāda, "although there may be a great increase in population on the surface of the earth, if the people are exactly in line with God consciousness ... such a burden on the earth is a source of pleasure for her."

Newspapers have become overpopulated, so to speak, with warnings about human overpopulation. Such warnings have been issued regularly for decades - even centuries - with consistently incorrect predictions. On the first Earth Day, Paul Ehrlich's 1968 bestseller, The Population Bomb, was widely quoted. He predicted that by 1985, the "population explosion" would lead to world famine, the death of the oceans, a reduction in life expectancy to 42 years, and the wasting of the Midwest into a vast desert. He was about as accurate as Malthus himself, the Englishman who, in 1798, predicted catastrophic food shortages that never came.

If the population is good, then no matter how numerous, they will be able to cooperate with each other peacefully and, with the blessings of God, receive ample resources from Mother Earth. A population of good character will not generate as much "vice and misery," and this is desirable for the health of the environment.

But could it be that we are running out of space? Walk through New York, Kolkata, or Hong Kong and experience the incredible crowding: surely there just isn't room for all these people. Yes, there are crowded places in the world. Yet leave these population centers, and we find a remarkably unpopulated planet.

Present oil crisis, like most other environmental problems can be traced to human vices, especially greed, leading to a vicious karmic cycle.

# Solution -7

## Zero Growth / Selective Growth Movements

*Learn to pause...or nothing worthwhile will catch up to you.*
*- Doug King*

## Speedometer vs Milestone
## Speed vs Direction

Growth for the sake of growth is the ideology of the cancer cell, says Edward Abbey.

World progress has increasingly come under scanner of late. The question being asked has three dimensions - first whether have we really made progress when progress is taken to mean overall well being and happiness of society. Secondly whether the progress has been in the right direction. While busy looking at the speedometer, have we missed the milestone. What would be the use of speeding along if the road we have taken is the wrong one. Thirdly, whether the so called progress is really required.

Some proponents of zero growth theory propose that faculty of human enjoyment is limited. A poor man may eat two meals but a millionaire can not eat 200 meals. A poor sleeps on a 6 feet couch, a tycoon can not occupy 60 feet space while lying. Our capacity to eat, sleep, mate etc. is limited. Therefore, so called economic development can not increase our enjoyment. Technically speaking, a poor man can sleep happily in his shack and a rich man may not get sleep in his plush apartment. Then who is enjoying more becomes a subjective

> *Now there is petrol problem. I see so many buses, and not a single man, one or two men. And for two men a big huge bus is being run, and so much petrol is consumed unnecessarily. I have seen. I went from Nairobi to London in a plane—only five passengers. Out of that, four passengers we were. Why? Why this nonsense? And there is petrol problem now. They are creating simply, the so-called advancement of civilization, creating problems, that's all.  - Srila Prabhupada*
> *(Room Conversation with Richard Webster, May 24, 1974, Rome)*

discussion. As per this theory, human happiness is independent of material facilities or accumulations. Developed countries have higher rate of suicides, stress and psychiatric cases as compared to the underdeveloped world.

In 2002 Australian liberal political theorist Clive Hamilton published a book about economics and politics by the name Growth Fetish which became a best-seller in Australia. The book has been the subject of much controversy, and has managed to infuriate commentators on both the left and right of the politico-economic debate.

The thesis of the book is that the policies of unfettered capitalism pursued by the west for the last 50 years has largely failed, since the underlying purpose of the creation of wealth is happiness, and Hamilton contends that people in general are no happier now than 50 years ago, despite the huge increase in personal wealth. In fact, he suggests that the reverse is true. He states that the pursuit of growth has become an addiction, in that it is seen as a universal magic cure for all of society's ills. Hamilton also proposes that the pursuit of growth has been at a tremendous cost in terms of the environment, erosion of democracy, and the values of society as a whole, as well as not delivering the hoped for increases in personal happiness. One result is that we, as a society, have become obsessed with materialism and consumerism. Hamilton's catchphrase "People buy things they don't want, with money they don't have, to impress people they don't like" neatly sums up his philosophy on consumerism.

Hamilton proposes that when a society has developed to the point at which the majority of people live reasonably comfortably, the pursuit of growth is pointless and should be curtailed. The surplus wealth could then be diverted into the essential infrastructure and to other nations that have not reached this level of wealth. Hamilton adapted the term Eudemonism to denote a political and economic

*To survive the chaos and crisis down the road, we can't continue our present oil-based, growth-crazy, throwaway economy.*
*. - V.B. Price*

model that does not depend on ever increasing and ultimately unsustainable levels of growth, but instead "promotes the full realisation of human potential through ... proper appreciation of the sources of well-being", among which he identifies social relationships, job satisfaction, religious belief for some, and above all a sense of meaning and purpose."

Hamilton relates the fetish for growth to a "development mentality", and to a neoliberal "instrumental value theory [which] maintains that, while humans are valuable in and of themselves, the non-human world is valuable only insofar as it contributes to the well-being of humans."

Clive Hamilton is the head of the Australia Institute, an independent think-tank. It is widely regarded as one of the very few viable left-leaning research centres in the country. Growth Fetish itself reflects many of the findings from the AI's report *Overconsumption in Australia*, which found that 62 per cent of Australians believe they cannot afford everything they need, even though in real terms their incomes have never been higher. The report also found that 83 per cent of people felt that society was "too materialistic", with too much emphasis on money and things, instead of what really matters. The Institute is also researching the growing phenomenon of downshifting, which Hamilton feels may be a response to the growth fetish, laying the foundation for a post-growth society.

When a doctor sees a patient's vital signs going off the chart, he knows it's time for emergency medication, perhaps too late. But what about a society? A chart of mankind's vital signs over the last thousand years would look like a patient going terminal. Take whatever indicator you like - environment, ethics, crime, energy consumption, $CO_2$ emissions - and graph it. The left three-fourths of the chart would be almost a horizontal line, followed by an almost vertical line covering the last 200 years. Is it time for remedial action?

We have come to think of growth as the Great Benefactor. Rapid growth has been the hallmark of the industrialized world, bestowing on its lucky denizens a standard of living unmatched in human history. Governments of every creed - capitalist, communist, Islamic, whatever - strive to promote ever greater economic growth. But should we be thinking of growth not as the Great Benefactor but as the Great Destroyer? Has growth become like a malignant cancer, devouring the very body which sustains it?

It may sound little radical but it's relevant today. According to Vedic conception, everyone has an allotted amount of happiness and distress which they are destined to experience during their lives. This quota of happiness and distress is fixed and cannot be changed by any material advancement. If the scientists eliminate one cause of suffering mother nature will find another way to inflict the suffering.

It is not possible to increase the amount of pleasure we experience by exploiting the resources of the planet. We get what we deserve as a result of our past karma. One person, who has good karma, may not work very hard and still lives a very comfortable life, whereas someone else, despite working very hard, can't manage two square meals.

In western countries we can see that despite so much advancement, they are no happier than their parents or grandparents. In fact, because of the dramatic increase in rape, incest, murder, violent crime, theft, stress and child abuse they are generally in a state of greater anxiety which is not at all conducive to happiness or peace of mind.

# Solution -8

## Religious Ethics Of Frugality, Thrift And Simplicity

*"One of the greatest challenges to contemporary religions is how to respond to the ecological crisis perpetuated by the enormous inroads of materialism and secularization in contemporary societies, especially those societies arising in or influenced by the modern West."*
– Dr. Mary Evelyn Tucker

## Religious Ethic Of Frugality, Thrift And Simplicity

Whether we are actively religious or not, religious belief permeates the very fabric of our existence. Namely, it influences, if not directly shapes, our personal, economic, social, ethical life and also our legal systems. It is then only logical to surmise that religion also influences how we, individually and collectively, view our role with regards to protecting our resources and environment.

What role does religion play in shaping our attitude towards the natural world? One answer was proposed in 1967 by UCLA History Professor Lynn White, Jr., who wrote an article entitled, "The Historical Roots of Our Ecological Crisis" (Science 155, 1967). In this article, he said that the Western world's attitudes towards nature were shaped by the Judeo-Christian tradition (he also included Islam and Marxism within this overall tradition). This tradition, White wrote, involved the concept of a world created solely for the benefit of man: "God planned all of creation explicitly for man's benefit and rule: no item in the physical creation had any purpose save to serve man's purposes." Along with this, Western Christianity separated humans from nature. In older religious traditions, humans were seen as part of nature, rather than the ruler of nature. And in animistic religions, there was believed to be a spirit in every tree, mountain or spring, and all had to be respected. In contrast with paganism and Eastern religions, Christianity "not only established a dualism of man and nature but also insisted that it is God's will that man exploit nature for his proper ends." White noted that Christianity was a complex faith, and different branches of it differ in their outlook. But in general, he proposed that Christianity, and Western civilization as a whole, held a view of nature that separated humans from the rest of the natural world, and encouraged exploitation of it for our own ends.

"Christianity," wrote White, "Not only established a dualism of man and nature but also insisted that it is God's will that man exploit nature for his proper ends." The emergence of Christianity, many, like White believe, marked the moment humans broke away from previously common held beliefs that all beings, all forms of life, including plants, had spirits (or souls).

"In Antiquity every tree, every spring, every stream, every hill had its own genius loci, its guardian spirit," he wrote. And Christianity changed all that, he believed. Man was created in God's image, Christians believed and notably Man was created at the end of Creation and humans therefore inherited the Earth. "By destroying pagan animism," White wrote. "Christianity made it possible to exploit nature in a mood of indifference to the feelings of natural objects."

There has been much discussion on Lynn White's articles but in general religion is meant to provide man with a philosophy of life and with prescribing practices that will assist him in living up to its teachings and in becoming a more spiritually conscious being.

Science has no "silver bullet" to fix the environmental crisis. Only a change of values could remedy this. Science is not capable of influencing large numbers of people to change their values; religion is the only force capable of doing this.

Religion is the only force that could bring about the change in values needed to reduce consumption and thus reduce environmental degradation. Environmental crisis is a spiritual one that demands a spiritual solution.

Of course, not every religious teaching is going to be helpful in this regard. Some contemporary manifestations of religion encourage material acquisition and de-emphasize contemplative spiritual practices.

*On Conservation*
*In our childhood we were taught by our parents that if a grain of rice falls on the floor, we must pick it up and touch it to our head to show respect. We were taught like this—how to see everything in relationship with Krishna. That is Krishna consciousness.*
*- Srila Prabhupada*

Religions need to be in dialogue with other disciplines (e.g., science, ethics, economics, education, public policy, gender) in seeking comprehensive solutions to both global and local environmental problems. Cultivation of spiritual sources of satisfaction, by prayer and meditation will reduce the burden on natural resources like oil as the urge to drive cars would be lessened.

The "What Would Jesus Drive?" ad campaign asking drivers to forgo gas-chugging SUVs drew both jeers and cheers. Voices of faith are taking stands on environmental issues with increasing boldness. National coalitions have been bolstering and diversifying their membership and now include Muslims, Jews, Buddhists and Hindus. Meanwhile, more local groups are organizing educational campaigns and lobbying legislators. Most activity focuses on fuel economy and global climate change, but in all faith traditions and all regions, groups are pressing issues they see as critical to being good stewards of the earth.

### Bishnoism

How religion influences our dealing with nature and natural resources can be seen from the example of Bishnoism. This is a sect

of vedic religion, originating in the desert of Rajasthan, India in the 15th Century. Bishnoism emphasizes love, peace and harmony not only among human beings but also with wild animals and trees. It teaches love, peace, kindness, simple life, honesty, compassion and forgiveness.

Living in inhospitable desert terrain, for centuries Bishnois have fiercely protected the trees and wild life in their areas to follow the teachings of their Guru Jambheshwarji. For Bishnois, caring for God's creation is their dharma or duty towards God. Time to time, their faith was tested by rulers, poachers and others, but Bishnois always protected the nature, even at the cost of their lives.

In 1730, 363 Bishnois were killed when they opposed cutting of Khejari trees. They hugged the trees and said, "sir santhe runkh rahe to bhi sasto jan", meaning if trees can be saved at the cost of our heads even then its a good deal. These trees were being cut at the order of the ruler, Raja Abhay Singh for firewood to burn lime stone to construct his palace.

These people will starve to feed other hungry creatures. Such is the awe and reverence these people have for God's creation that ladies will breast feed an orphan calf. Over the years, hundreds have died in their attempt to save nature and wildlife. This presents a sharp contrast to western mindset which shows utter disregard for nature, environment and other life forms.

Communities like Bishnois, once scorned for being backward, have a valuable lesson to teach us.

## Jainism

The Jain tradition has existed in tandem with Vedic tradition in India since at least 800 BC. Jains developed their own sacred texts and follow the authority of itinerant monks and nuns who wander throughout India preaching the essential principles and practices of the faith. Jainism holds some interesting potential for ecological thinking, though its final goal transcends earthly concerns i.e., ascending to the realm above earth and heaven.

At the core of Jain faith lies five vows that dictate the daily life of Jains. These five vows, which inspired and influenced Mahatma Gandhi, are nonviolence (ahimsa), truthfulness (satya), not stealing (asteya), sexual restraint (brahmacarya), and nonpossession (aparigraha). One adheres to these vows in order to minimize harm to all possible life-forms. For observant Jains, to hurt any being results in the thickening of one's karma, obstructing advancement toward liberation. To reduce karma and prevent its further accrual, Jains avoid activities associated with violence and follow a vegetarian diet. The advanced monks and nuns will sweep their path to avoid harming insects and also work at not harming even 'one-sensed' beings such as bacteria.

The worldview of the Jains might be characterized as a biocosmology. Due to their perception of the "livingness" of the world, Jains hold an affinity for the ideals of the environmental movement. The Jain vows can easily be reinterpreted in an ecological fashion. The practice of nonviolence in the Jain context fosters an attitude of respect for all life-forms. The observance of truthfulness prompts an investigation of the interrelatedness of things; a truthful person cannot easily dismiss the suffering caused by uncontrolled waste. The vow of not stealing can be used to reflect on the world's limited resources and prompt one to think of the needs of future generations. Sexual restraint might help minimize population growth. The discipline of nonpossession gives one pause to think twice before indulging in the acquisition of material goods, one of the root causes of current resource crunch. The monks and nuns, due to the heightened nature of their daily spiritual practice, leave little or no imprint on the broader ecological system.

The Jains are well-suited to reconsider their tradition in an ecological light, particularly because of their history of advocacy against meat eating and animal sacrifice, as well as their success at developing business areas that avoid overt violence. Thus directly and indirectly these qualities can help preservation of finite natural resources like oil.

### Bhagavata Dharma

Vedic tradition is popularly known as Sanatana dharma or Bhagavata dharma. Srila Prabhupada sums up vedic ecology in the following words:

"The human being is the elder brother of all other living beings. He is endowed with intelligence more powerful than animals for realizing the course of nature and the indications of the Almighty Father. Human civilization should depend on the production of mother nature without artificially attempting economic development to turn the world into a chaos of greed and power only for the purpose of artificial luxuries and sense gratification." *(Srimad Bhagavatam 1.10.4 purport)*

Bhagavata Dharma considers earth to be one of the seven mothers. Other six mothers are : (1) the real mother, (2) the wife of the spiritual master, (3) the wife of the priest, (4) the wife of the king, (5) the cow, (6) the nurse.

Material nature is considered one of the energies of the Supreme Lord. Bhagavad-gita(6.30) mentions "For one who sees Me everywhere and who sees everything in Me, I am never lost nor is he ever lost to Me." So a follower of Bhagavata dharma perceives the presence of the Supreme Lord everywhere and in everything. As per 'isavasyam idam sarvam' concept, everything in this universe is the property of the Supreme Lord and all living beings have been allotted resources of material nature to maintain themselves. We should take our quota and should not encroach upon what is meant to be others' quota. Therefore in Bhagavata Dharma, there is no scope for hurting environment or wasting natural resources. A householder is responsible for maintenance of not only his own family members but all other living beings.

Srimad Bhagavatam(7-14-9) explains:

mrgostra-khara-markakhu-

sarisrp khaga-maksikah

atmanah putravat pasyet

tair esam antaram kiyat

One should treat animals such as deer, camels, asses, monkeys, mice, snakes, birds and flies exactly like one's own son. How little difference there actually is between children and these innocent animals!

This is in sharp contrast to the baseless western idea that animals have no soul. Those who propagate such idea and kill poor animals mercilessly in fact do not have soul. Bhagavata vision of equality is explained in Gita (5.18)

vidya-vinaya-sampanne brahmane gavi hastini

suni caiva sva-pake ca panditah sama-darsinah

The humble sages, by virtue of true knowledge, see with equal vision a learned and gentle brahmana, a cow, an elephant, a dog and a dog-eater.

People today have no regard for life, no regard for nature and her precious gifts. But a Bhagavata follower can never lead a wasteful life, squandering away precious natural resources like oil.

# Solution -9

## Mantra Meditation
## For
## Non-material Satisfaction and Enrichment

## Overcoming Overconsumption And Thereby Fuel Shortages By Mantra Meditation

S cience, in spite of all the developments, has failed to comprehend the reality of matter, reality of consciousness and the unity of the two. The East, where the knowledge of reality exists, has gone subservient to the west and the reality of nature or the secret of Karmic cycle has remained a paradox to humanity and science.

Science has advanced based on the Knowledge of matter for the last one thousand years. End of it should be a transformation into the knowledge of the soul and Supreme Soul and its superiority and control over matter. In other words the knowledge or the wisdom of life, that existed before but was forgotten by humanity. Only the knowledge of interrelationship matter and soul and control of both by the Supreme Spirit is the answer to environmental and resource anomalies.

Lacking spiritual understanding, people instinctively pursue material gratifications and possessions, thus fueling the overconsuming economy that overwhelms the world with shortages of all kinds, including gasoline one.

In the glare of industrialization, we have lost all connection with the world of spirit. This is a sentiment recognised nearly a century ago by the Austrian poet, Rainer Maria Rilke: "The whole 'spirit world', death, all those things that are so closely akin to us, have by daily parrying been so crowded out by life that the senses with which we could have grasped them are atrophied. To say nothing of God."

Through the application of ancient sacred sounds, we are infused with energy that dissolves our difficulties and improves our lives. Over time our karma becomes exhausted, and we become spiritually free of any worldly bondage.

Great spiritual teachers in the Vedic tradition therefore advise that we reconnect ourselves with God, the supreme person to be completely free from karma and get non-material satisfaction. This can be done through devotional spiritual practices.

Mantra meditation is a radical re-envisioning of ourselves, our lives and our ability to create the future we desire. The principles of mantra meditation are based on a classical Eastern model of how the universe operates, and our place and purpose in it.

The Vedas explain that powerful spiritual energies can be generated by yoga, meditation, and the chanting of mantras. In the present age, the chanting of mantras is particularly effective. When properly chanted, the combinations of sounds in mantras release their energies. Bible also proclaims, "In the beginning was the Word, and the Word was with God, and the Word was God."

The most powerful mantras, according to the Vedas, are those composed of names of God, such as the Hare Krishna mantra: Hare Krishna Hare Krishna Krishna Krishna Hare Hare / Hare Rāma, Hare Rāma, Rāma Rāma, Hare Hare. The Vedas teach that God's name, being nondifferent from God Himself, is supremely potent. The chanting of the Hare Krishna mantra is especially effective when people chant it aloud together.

If one chants the holy names of God like Hare Krishna mantra, he will realize that he is not this material body and does not belong to this material world. Thus sacred sound syllables can repair our

*"So, if you chant this Hare Krsna, then you will very easily understand that you are not this body; you are spirit soul, aham brahmasmi: 'I am Brahman." Ceto-darpana-marjanam. That is the first installment, you'll understand. For self-realization, there are so many different processes. There are mystic yogis. There are philosophical speculators. There are karmis. There are jnanis. And... But this process, immediately you will realize that you are not this body. You are not matter; you are spirit soul. Ceto-darpana-marjanam. And, as soon as you understand that you are not this body, you are not matter, that you are spirit soul, then immediately you become joyful."*
*~Srila Prabhupada.*

estranged relationship with nature. This ancient practice can change the way we connect with our environment.

## Chanting The Holy Name Is The Ecology Of The Mind.

The holy name is ever fresh, beautiful and joyful. Chanting provides non-material happiness, reducing the need for material enjoyment. The mind completely lacking in spirituality only knows the world through prices, figures and statistics. A mind saturated with non-material satisfaction does not require so many material objects for sense gratification. Any one who chants the Holy Name will be satisfied with a simple life and a simple life is always easy on finite natural resources like oil.

# Solution -10

## Ox Power

*What is the use of car? If you locate yourself (get localized) to get everything, your necessity, then where is the use of car? If you require car, you have a bullock cart. That's all. Why should you hanker after petrol, mobil, oil, machine, this, that, so many things. Why?*
*–Srila Prabhupada (Conversation, October 5, 1975, Mauritius)*

# Ox Power

Reliance on fossil fuels has largely developed in the last 200 years. Before that, most energy was renewable – animal and human muscle, wood, some wind and water power. The harnessing of new sources of energy, especially coal, about 250 years ago was crucial to the industrial revolution and all that followed.

For millennia animals have been harnessed to pull carts, carry loads, transport people, haul water, trash harvests, plough, puddle and weed crop fields etc. **Even today, more than half the world's population depends on animal power** for much of its energy. Draught animals operate on more than 50% of the planet's cultivated areas. In the mid 1990s work by draught animals was estimated to be equivalent to a fossil fuel replacement value of US\$ 60 billion. Estimates of the number of animals used for power applications range from 300 million upwards. Oxen  are the most frequently used animals and ploughing is the most common function. Almost all species of domestic quadruped are used, however, in a variety of agricultural and transport roles. In agriculture positive effects are seen to be higher crop output, better returns to labour, increased cash income and improved food security.

Despite motorization on all fronts the use of animals is still often more economic than the use of machinery and vehicles, especially in small scale agriculture and in remote areas. Animals are produced and maintained locally and don't require the infrastructure needed for motorization. Where the value of machinery needs to be depreciated over time, that of animals can appreciate because of growth.

Ox power represents a sustainable and renewable resource of energy.

In terms of agriculture, ox power creates a lighter footprint on the earth than a tractor, which tends to compact the soil. Also in terms of the environment, it takes far less resources to produce a team of oxen than a tractor. How many mining operations and how many factories does it require to produce one tractor? How many drilling and refining operations does it take to fuel it? The "factory" that produces an ox is a cow. For "fuel" the oxen can eat grass and grain which they themselves produce.

And, we should not underestimate the level of benefit that oxen can provide. With the exception of the cultures of the Americas, practically every materially advanced civilization before the crusades – including China, India, the Middle East, North Africa and Europe – relied on oxen as the engine of agriculture and also for local transport, grinding of grains and even building. Many of the great projects of ancient times were all accomplished without the incredible level of pollution it would take to recreate such structures today.

Srila Prabhupada advises, "Petrol is required for long-distance transport, but if you are localized, there is no question of such transport. You don't require petrol.... The oxen will solve the problem of transport."

The tractor is a real sore point in agriculture. Tractors are expensive to operate. This expense partly explains why 30,000 to 50,000 small farms collapse every year in US. But ox power, though slower, is far more efficient.

For small farms, oxen do better than tractors. They require no gasoline, cost far less than tractors to maintain, provide free fertilizer, preserve precious topsoil, and don't foul the atmosphere with carbon monoxide. Bovine waste, when mixed in the traditional way with straw, is the world's best fertilizer. And when the animal dies, its skin can be processed into leather.

Turning to Gandhi for inspiration, we find that a key requirement for building peace is to provide full employment by emphasizing localized production for localized markets. Gandhi stressed that everything which can be produced locally should be, even if the local economy is less efficient at its production. Since time immemorial, human cultures have lived with and protected cows. Cows have provided many essential services to humanity for very little maintenance. They're an inseparable part of God's efficient system for human civilization. Today people employ them in agriculture in India and many other so-called developing countries.

Dr. Vandana Shiva, an ecologist, comments on India's recent cattle policy while calling it a policy of ecocide of indigenous cattle breeds and a policy of genocide for India's small farmers: "The traditional approach to livestock is based on diversity, decentralisation, sustainability and equity. Our cattle are not just milk machines or meat machines. They are sentient beings who serve human communities through their multidimensional role in agriculture."

"On the other hand," continues Shiva, " externally driven projects, programmes and policies emerging ftom industrial societies treat cattle as one-dimensional machines which are maintained with capital intensive and environmentally intensive inputs and which provide a single output - either milk or meat. Polices based on this approach are characterised by monocultures, concentration and centralisation, non-sustainability and inequality."

Thus, whether we like it or not, when fossil fuels bid us good bye, we will have to revert back to bull power for fulfilling some of our energy requirements.

# Solution -11

## Deurbanize - Promote Rural Living

*Better give up city. Make Vrindavan, like this. City life is abominable. If you don't live in the city, you don't require petrol, motor car. ~Srila Prabhupada*
*(Room conversation, August 1, 1975, New Orleans)*

Urba is the Latin for town. Urban is anything of or about a town. Suburb is an outlying part of an urban area partway between urban and rural areas. Suburbs are usually associated with being dormatory (housing) areas (with green spaces) as support for the more central urban area.

In order to define and distinguish movements and divisions of populations within a given territory, there are four popular concepts that have been identified - 'urbanization', 'suburbanization', 'de-urbanization' and 're-urbanization'. Typologically speaking, during initial phase of 'urbanization', there occurs an import of rural traditions into the city centers. Deurbanisation, which is just opposite of urbanization, is defined as breakdown of urban areas into non-urban areas. Urban, suburban and rural areas can be defined by population density. So "deurbanisation" is the reverse process of turning heavy population density areas into lower density areas.

The world is rapidly urbanizing. The UN Population Division estimates that by 2017 half the world's population will be urban. As indicated by Redman and Jones (2004):

"Cities occupy 4% or less of the world's terrestrial surface, yet they are home to almost half the global population, consume close to three-quarters of the world's natural resources, and generate three-quarters of its pollution and wastes. Moreover, the UN estimates that virtually all net global population and economic growth over the next 30 years will occur in cities, leading to a doubling of current populations. This growth will require unprecedented investment in new infrastructure and create extraordinary challenges for political and social institutions."

Urbanization is particularly rapid in the developing world, where major economic restructuring in countries like China, and the lack of rural employment opportunities in many African and Asian countries, is provoking an exodus from rural areas to towns and cities. Although much of the focus has been on the growth, infrastructural and environmental problems of megacities (those over 10 million in population), the reality is that much urbanization is projected to take place in the small to medium sized cities (e.g., former provincial towns), and not just large cities.

This poses numerous challenges like environment, health, conversion of cropland, forest and wetlands to urban "built up" areas; inadequate provision of improved water and sanitation, particularly in slums and ghettos; waste removal; and air pollutant emissions from industry and transportation.

We can draw parallels to the economic decline of Rome with the impending economic decline of America. Another parallel could be drawn to take in the issue of deurbanization. Rome was preeminently an urban empire. It's power was founded on cities, particularly Rome itself. But when the price of slaves rose, labor shortages in the country caused an exodus from the cities to the country. Our modern cities conceivably could be facing a similar exodus. The first major causality of high oil prices will be the car. And that would include gas/diesel powered trucks. In any case, transporting food will become increasingly expensive. And driving several miles to a big supermarket will of course also present problems. Many people will want to grow their own food, or at least be close to places where food is readily available at decent prices. This would mean a flight to the country, as cities become more and more expensive to maintain.

## Recent Trends In Deurbanization

1. Linked to recent industrial changes: companies have moved to lower cost areas.
2. Technological change - e.g. Internet etc - people can work from home.
3. Idealistic views of the idyllic countryside where there are less social problems such as crime, muggings and drugs etc.
4. Better quality of life.

When we talk of deurbanization in the context of post-petroleum civilization, we do not mean suburbanization which is more wasteful than urbanization itself. Example of suburbanization can be seen when in the 1960s, American factories began to be systematically removed from the central urban cores. This drove the development of suburban neighborhood design - allowing the urban elites to distance themselves both physically and socially from the working class. A precursor again to 'de-urbansiation.'

### Resource Extravagance of Cities in Comparison to Rural Areas.

The urbanites inflict ecological damage not only on the urban areas they cement over but also on the hinterland from which they draw resources. Cities are centres of consumption and extremely resource-intensive. The large scale, centralized systems they require are almost without exception more stressful to the environment than small-scale, diversified, locally adapted production. Food and water, building materials, and energy must all be transported great distances via energy consuming infrastructures; their concentrated wastes must be hauled away in trucks and barges or incinerated at great cost to the environment.

### Urbanites Divorced from the Soil.

The reason urbanites cause so much ecological devastation is because living in cities encourages people to keep demanding more and more commodities without appreciating the impacts of these demands on nature. Urbanites insist on cheap, flawless agricultural products and this exerts considerable pressure on farmers to use monocultural methods of agricultural production. By the very fact that they are locked away from the Earth in an artificial environment, urbanites lose sight of the Planet as a living entity with whom they must maintain an organic reciprocity. Destruction of nature will increase in scale as cities become even more extensive.

### The Shortage Of Fossil Fuels Will Lead To The Decline Of The City

Some feel that in the long term, deurbanization is inevitable because as fossil fuels become scarcer, industrialized agriculture will not be able to feed the increasing number of urbanites, Eventually the proportion of farm to city population will have to reverse itself if humanity has to survive. Labour intensive organic farming cannot support the concentrated urban population centres that have been built up during the high energy, fossil fuel age.

## Village Life

In a post-industrial world, the self-sufficient agricultural village, rather than the urban factory or rural factory farm, will be the primary economic unit.

The use of the term 'village' is very significant, and comes with a definition. A village is human-scale: it is large enough so that all the needs of its inhabitants can be met, with complete specialization of tasks, but not so large that there are anonymous people; in a village, everybody is known, and strangers are instantly recognized and assessed. The actual size depends on the 'carrying capacity' of the surrounding ecosystem. Settlements smaller than villages are called 'hamlets'; in a hamlet there are not enough people for a diversity of specialization of tasks, so culture is limited to necessities. A 'town' has enough people so that not everyone is known, and suspicions develop concerning the motivations of strangers. A town has outgrown the carrying capacity of the surrounding landscape, so a polarity of interests develops over the allotment and use of ever-scarcer resources. A 'city' is a blight on the landscape, and results in the complete impoverishment of the natural ecology. Strangers are numerous and a general feeling of distrust, anxiety, and fear prevails, leading to isolation and alienation. Civilization, the culture of cities, is an abstract human construction, and is completely divorced from natural processes. The goal of such a civilization is to maximize and concentrate arbitrary power, and this eventually leads to its own self-annihilation, in every case. With this review of human settlement patterns, it becomes obvious that the optimum blend of diversity and sustainability occurs at the village scale. It is there that humans have the opportunity to achieve maximum self-realization.

If Homo-sapiens want a sustainable future and survival into the indefinite future, they must reorganize themselves at the village scale.

In the words of E. Christopher Mare, "The global economy is so insidious — it has forced itself into every nook and crevice of the globe. Like a toxic slime, it has covered everything, and adversely affected the lives of everybody, extracting the life out of whole communities. What is truly tragic is that it has infected the minds of the people as well, so that they are not even aware of their debilitating collusion."

Once a local economy becomes dependent on the global economic system, it has entered the murky, life-draining realm of un-sustainability, because the global economy itself is not sustainable.

Vedic system of Village organization is based on a concept called Varnashrama. Under this system, our life span (assumed to be 100 years) is divided into four 'ashramas' or life divisions known as brahmachari (celibate student), grihastha (married householder), vanaprastha (retired) and Sannyasi (renounced). Similarly whole society is divided into four classes. Srila Prabhupada explains this, "The brahmanas study transcendental literature such as Bhagavad-gita and the Upanishads. And they lecture and instruct, as well as worship the Deity in the temple. They should have ideal character, and the other classes provide food and shelter out of appreciation for their guidance. The ksatriyas, taking advice from the brahmanas, govern the village and apportion land to the vaisyas, who use the land to produce grains, fruits, and vegetables and to raise cows for milk. The vaisyas give twenty-five percent of their produce or earnings to the ksatriyas, who utilize it for village projects. The sudras - the artisans and craftspeople - assist the other three classes."

## A Case Study

In early 70s, Hare Krishnas founded a farm community near New Orleans, named New Talavan. Hare Krishna guru, Srila Prabhupada advised the community to strive for self-sufficiency and grow their own grain, fruit, and vegetables. They should keep a few cows for milk-which they could then turn into yogurt, butter, and fresh natural cheese. They should use oxen to plow the fields and for local transport, grow sugar cane for sweetener and grow castor beans and use the oil to burn in lamps.

They should grow cotton, spin it into thread, and weave their own cloth on handlooms. For building materials, they should use logs and bricks. Finally, Srila Prabhupāda encouraged the residents of New Tālavan to build a magnificent temple at the center of the community, to provide spiritual nourishment.

Later that year, Srila Prabhupāda visited New Talavan and gave additional advice about how to organize the community. "Avoid machines. Keep everyone employed as a brahmana [teacher], ksatriya [administrator], vaisya [farm owner or merchant], or sudra [laborer]. Nobody should sit idle." He was explaining the Vedic social system, with its natural divisions or classes that allow people to make the most of their special aptitudes and inclinations.

# Solution -12

## Deglobalize
## Decentralize
## Localize

*No mention of there being too many people and too many people with large appetites for energy. Time to conserve energy. Move closer to your work and shopping. Move where you can walk or bicycle to whereever you need to go. Go from a multi-car family to a one car family and save money on gas, car insurance and the car itself. And let's get away from globalization and back to bioregionlism. Take the farms away from the corporations and let the local people go back to farming.*
*- Karen Gaia*

## Globalization Vs Localization

Supplying enough energy on a reliable basis, at prices that won't totter world economic growth, is emerging as a challenge with repercussions that are hard to predict. For oil and gas companies and others in the energy business this means new opportunities but also serious risks. Inexorably, energy demand is growing — not only in the developed economies of Europe, Japan and North America, but in developing nations as well. In fact, the fastest demand growth is in China, India and other emerging markets. From one side of the globe to the other, modern and modernizing societies need more fuel.

But the places with the greatest demand can't supply their own needs. Over the next few decades, oil and gas production in the North Sea, North America and China are expected to fall, or rise too little to keep pace with demand.

In conclusion we will have to move towards a much more "in situ" economy! Globalization feeds on cheap energy.

### Deglobalization

Economic localization is the process by which a region, county, city, or even neighborhood frees itself from an unhealthy dependence on the global economy and looks inward to produce a significant portion of the goods, services, food, and energy it consumes from its local endowment of financial, natural, and human capital. Economic localization brings production of goods and services closer to their point of consumption, reducing the need to rely on long supply chains and distant markets so that communities and regions can, for the most part, provision themselves.

While it is certainly not possible to produce every kind of goods and services locally, economic localization seeks to restore an efficient balance between local production and imports that fully accounts for the social and environmental costs neglected by free trade agreements. Local production strengthens the local economy, creates worthwhile jobs, and increases local self-reliance. Refocusing the economy locally will necessarily revitalize the community, increasing intimacy, cooperation, and support for local culture and a sense of place.

Local production for local consumption also reduces the need to ship materials and products large distances. Reducing transport lowers $CO_2$ and other pollutant emissions and reduces dependence on burning fossil fuels. This is particularly important considering that the transportation sector is the largest emitter of $CO_2$ and a near consensus of scientists believe that global $CO_2$ emissions need to be reduced 60-80% to have a possibility of stabilizing the climate. This will not occur without reconfiguring our economies and cities for much less transport and energy consumption.

In light of widely predicted oil shortfalls in the years and decades to come, localization appears to be inevitable. According to the U.S. Department of Energy-funded "Hirsch Report," it would take two decades for an orderly transition off our oil-centered transportation system. And this report did not address all the other things for which we use oil such as plastics, synthetic fabrics, artificial rubbers, fertilizers, and pesticides. As oil peaks and becomes increasingly less available and prices rise, there will be an unyielding economic push towards localization.

*Prabhupada : Bus...containing three passengers, wasting petrol. Similarly, hundreds and thousands and millions of cars and buses are running all over the world, simply wasting petrol.*
*Bhagavan : When there was the oil crisis in the United States, they were giving reports how some person would go in his car, go ten miles in a big car to buy one pack of cigarettes.*
*Prabhupada : Stick to your own place and grow your food. There is no question of transport. Little transport is required, that bullock cart. Krishna was being carried on bullock cart. There is no use of petrol. Use simply the bull. They are already there. Utilize them. (Morning walk, May 25, 1974, Rome)*

Ever-rising gasoline prices will force many long-distance commuters to relocate closer to their jobs, increasing demand for urban housing. This will accelerate the regentrification of urban areas and the corresponding displacement of low-income urban residents to slums and older and outer suburbs. While the suburbs may initially seem attractive for some former urban residents, they will lose their luster as gasoline prices continue rising and wealthier residents abandon suburbia for the city. Further, as energy and fuel prices continue to rise, urban and suburban families living from paycheck to paycheck (or on the margins) and transport-intensive businesses will be increasingly stressed, if not destitute. Without swift response and action, the middle class will be next to feel the pain of utter dependence on a dwindling resource and inadequate preparation for the transition to an intensely local, post-petroleum future.

Therefore, choosing to progress now to a more localized economy will be a wise move, and less painful if it is planned for and managed well in advance of world peak oil production. The obvious starting point is an assessment of what each community already possesses both in natural and human resources and their abilities to produce and store food, energy, water and essential goods, with the aim to integrate these efforts into a parallel public infrastructure that can serve as a safety net when we hit hard times.

Turning to Gandhi for inspiration, we find that a key requirement for building peace is to provide full employment by emphasizing localized production for localized markets. Gandhi stressed that everything which can be produced locally should be, even if the local economy is less efficient at its production.

### Oil Crisis Solution - Localization

Following is a conversation with His Divine Grace A.C. Bhaktivedanta Swami Srila Prabhupada on the issue of oil crisis.

*Prabhupada* : Yes. We are going to solve all problems. Let us have some preliminary discussion, how we are going to solve.

*Bhagavan* : The biggest problem now is that they have built up a type of society in which their needs are all coming from petrol energy.

To produce what they need today is all coming from this petrol energy...

*Prabhupada* : Yes, yes.

*Bhagavan* : ...which they are importing basically from the Saudi Arabian countries.

*Prabhupada*: Yes.

*Bhagavan:* Now, recently, in the last war in the Middle East, Saudi Arabians raised the price of the oil over double now, I think, as a pressure to the Western countries to do things in their favor. Now they realized that the market for oil is in such great demand that they don't have to lower the price after the war, but they are going to keep the price. And actually the price is still increasing. So this is causing inflation.

*Prabhupada :* So this problem will be solved as soon as we are localized. Petrol is required for transport, but if you are localized, there is no question of transport. You don't require petrol. Suppose in New Vrindaban, we stay, we don't go anywhere. Then where is the need of petrol?

*Bhagavan* : Petrol they also use for heating. And electricity.

*Prabhupada:* No, heating. Heating we can be done by wood. By nature.

*Dhananjaya:* I remember, Srila Prabhupāda, you were saying that all we require is some oxen, and the oxen can carry.

*Bhagavan :* Yes. The oxen will solve the problem of transport. That bullock cart. Just like Krishna, when He was transferred from Gokula to Nandagrama, so they took all the bullock carts, and within a few hours they transported them, the whole thing, their luggage, family member, everything.

*Bhagavan :* How far can a bullock cart travel in one day?

*Prabhupada :* At least ten miles, very easily, very easily. And maximum he can travel fifteen miles, twenty miles. But when we are localized, we don't require to go beyond ten miles, five miles. Because we have created a rubbish civilization, therefore one is required to go fifty miles for earning bread, even hundred miles sometimes.

*Dhananjaya :* Like in Los Angeles.

*Prabhupada:* Why Los Angeles? Everywhere. In New York they are coming from hundred miles. From the other side of the island. First ferry steamer, then bus, then so on, so on. Three hours, four hours, they spend for transport.

*Satsvarapa :* Is this an ideal solution or a practical one?

*Prabhupada :* This is practical.

*Satsvarūpa :* Because sometimes we say that actually we cannot change the course of the...

*Prabhupada :* No, no. Our society will be ideal by practical application.

*Satsvarupa :* Not that we dictate to the... Not that we are going to force everyone.

*Prabhupada:* No, we are not going to force anyone. "Our mode of living is like this. If you like you can adopt." Just like we chant Hare Krishna mantra. So we are not forcing anyone that "You also, you must chant." No. We live like this.

*Dhananjaya:* So in fact, Srila Prabhupāda, we should start using bullock carts.

*Prabhupada:* Yes. No, first of all you start the community project, as we have already started in New Vrindaban. Make this perfect.

*Devotee :* There was a big meeting of scientists in Stockholm, Sweden, and they talked that if humanity don't begin to live in a localized way like you say, in fifty years will be no more source of production.

*Bhagavan :* Another important thing, there are three uses of petrol, or four. One is the transportation, one is heat, another is electricity, and a fourth is they use it to manufacture so many products. So what if someone asks.

*Prabhupada :* No, you go on with your product. You have created the problem, you go on with your problem. But we live like this. If you like, you can adopt.

*(May 27, 1974, Morning walk, Rome)*

# Solution -13

## Revival

## of

## Preindustrial Features of Life

*"Modern economy is a fire-breathing vampire of petroleum which is slowly cooking our planet."*
**-Thio Bode**
*Executive Director of Greenpeace*

There are still parts of the world where people live a pre-industrial life. For example take the case of Indonesian Borneo. Daily life in Borneo's upcountry is usually pleasantly dull, as chickens scratch around, the women fan rice on mats to dry it, thunderstorms roll through, the sun dries the muddy paths, flowers riot into bloom, and it all starts over again the next day. Pastoral Mongolia partially fits the category, too, with its world revolving around camels and sheep rather than rice and bananas.

Preindustrial life was easy on resources - both human and natural. Before capitalism, most people did not work very long hours at all. The tempo of life was slow, even leisurely; the pace of work relaxed.

Shifting away from this 'primitive' life has cost us a lot. In a vast number of ways and places, the biosphere of this planet has undergone a great deal of damage. Parts of the environment have been rendered uninhabitable through toxic wastes and nuclear power plant disasters, while systemic pollution, ozone holes, global warming and other disasters are increasingly tearing the fabric on which all life depends. That such damage is wrought overwhelmingly by corporations in a competitive international market economy has never been clearer, while the need to replace the existing society with a society with at least a few preindustrial features, has never been more urgent.

Modernization, the replacement of machines for muscle, is a universal social solvent. Even when resisted by traditional leaders, modernization erodes established social, economic patterns, and threatens ecosystems.

*Within my own life, I read all the beloved novels by lamps of vegetable oil; I saw the Standard Oil invading my own village, I saw gas lamps in the Chinese shops in Shanghai; and I saw their elimination by electric lights.*
*-Hu Shih*

Peasants and tribal members ultimately succumb to mechanisms yielding enhanced productivity. They rapidly scrap traditional practices in favor of those more materially productive. But in the long run, all this progress comes with a heavy price tag.

Our ancestors may not have been rich, but they had an abundance of leisure. When capitalism raised their incomes, it also took away their time. Indeed, there is good reason to believe that working hours from the mid-nineteenth century onwards constitute the most prodigious work effort in the entire history of humankind.

Consider a typical working day in the medieval period in Europe. It stretched from dawn to dusk but work was intermittent - called to a halt for breakfast, lunch, the customary afternoon nap, and dinner. Depending on time and place, there were also midmorning and midafternoon refreshment breaks. These rest periods were the traditional rights of laborers, which they enjoyed even during peak harvest times. During slack periods, which accounted for a large part of the year, adherence to regular working hours was not usual. According to Oxford Professor James E. Thorold Rogers, the medieval workday was not more than eight hours.

The contrast between capitalist and precapitalist work patterns is most striking in respect to the working year. The medieval calendar was filled with holidays. Official - that is, church - holidays included not only long "vacations" at Christmas, Easter, and midsummer but also numerous saints' and rest days. These were spent both in sober churchgoing and in feasting. In addition to official celebrations, there were often weeks' worth of ales -- to mark important life events. All told, holiday leisure time in medieval England took up probably about one-third of the year. And the English were apparently working harder than their neighbors. The ancient régime in France is reported to have guaranteed fifty-two Sundays, ninety rest days, and thirty-eight holidays. In Spain, travelers noted that holidays totaled five months per year. The peasant's free time extended beyond officially sanctioned holidays. A thirteenth-century study finds that whole peasant families did not put in more than 150 days per year on their land. Manorial records from fourteenth-century England indicate an extremely short working year - 175 days.

An interesting observation of pre-industrialized India was made by Lord MCLau (a British officer), on February 2, 1835 : "I have traveled across the length and breath of India and I have not seen one person who is a beggar, who is a thief, such wealth I have seen in this country, such high moral values, people of such caliber (of noble character), that I do not think we would ever conquer this country...........unless we break the very backbone of this nation which is her spiritual and cultural heritage."

Each era of human civilization is marked with issues but industrial era is phenomenal as far as the scale of destruction is concerned, destruction of lives, destruction of environment and destruction of natural resources.

We must do much more research on sustainable energy, economic planning, and community planning. We shall not give up our knowledge of electronics, quantum theory, and higher math, but we had better begin to salvage what we can of the ancient wisdom that we will need to tread lightly on the earth like older civilizations living in harmony with nature. Thus, we had better begin trying to learn from the few surviving patches of older cultures. We must begin to treat them as valuable endangered resources. Clearly, our lives might be enhanced by scientific knowledge, but we had better stop using it to subdue Nature rather than to create a partnership with her.

# Solution -14

## Sustainable Small Cities
## Ecocities or Ecopolis

*Mexico City, before the Spanish invaders came, was a floating city with 4-storey constructions built on rafts of living vegetation and had about 50,000 population, a nice population target for restructured, low-energy metropolis.*
*-Andrew McKillop*

Permanently increasing oil prices and at some point, the inability to obtain fossil fuels at any price will make much of our urban landscape more or less untenable. Politics and the economy will have to become more locally-organized. Since peak oil has to do with the amount of energy available to run virtually every system important to civilization, every sector of local society, economy, and culture will be affected in one way or another.

In a post-industrial world, there would still be cities, although they would be much smaller than those of today. They might resemble the cities of Renaissance Europe. Florence, for example, at the height of it cultural ascendancy, had only 35 to 45 thousand inhabitants.

A sustainable city, ecocity or ecopolis is an entire city dedicated to minimizing the required inputs (energy, water and food) and its waste output (heat, air pollution as co2, methane, and water pollution). Richard Register first coined the term "ecocity" in his 1987 book, Ecocity Berkeley: Building cities for a healthy future. Another leading figure who envisioned the sustainable city was architect Paul F. Downton, who later founded the company Ecopolis Architects.

Vedic literatures make mention of many beautiful cities in ancient India, like Mathura and Dwaraka which were full of gardens and artificial ponds.

A sustainable city can feed and power itself with minimal reliance on the surrounding countryside and creates the smallest possible ecological footprint for its residents. This results in a city that is friendly to the surrounding environment in terms of pollution, land use, and alleviation of global warming. It is estimated that by 2017, over half of the world's population will live in urban areas and this provides both challenges and opportunities for environmentally-conscious developers.

## Methodology

These ecological cities can be achieved though various means, such as:

-Different agricultural systems such as agricultural plots within the city (suburbs or centre). This is to reduce the distance food has to travel from field to fork. Practical work out of this may be done by either small scale/private farming plots or through larger scale agriculture (eg farmscrapers).

-Renewable energy sources, such as wind turbines, solar panels, or bio-gas created from sewage. Cities provide economies of scale that make such energy sources viable.

-Various methods to reduce the need for air conditioning (a massive energy demand), such as low lying buildings that allow air to circulate, natural ventilation systems, an increase in water features and green spaces equaling at least 20% of the city's surface. This counters the environmental heating caused by factors such as an abundance of concrete and asphalt, which can heat city areas by up to 10 degrees Celsius during the evening.

-Improved public transport and an increase in pedestrianisation to reduce car emissions. This requires a radically different approach to city planning, with integrated business, industrial, and residential zones. Roads may be designed to make driving difficult. Such induced walking would improve people's health and help preserve gas.

-Optimal building density to make public transport viable but avoid the creation of urban heat islands.

-Solutions to decrease urban sprawl, by seeking new ways of to allow people to live closer to the workspace. Since the workplace tends to be in the city, downtown, or urban center, they are seeking a way to increase density by changing the antiquated attitudes many suburbanites have towards inner-city areas. One of the new ways is on how this is achieved is by solutions worked out by the Smart Growth Movement.

-Green roofs

-Green transport

-Sustainable urban drainage systems or SUDS

-Xeriscaping - garden and landscape design for water conservation

## Moratorium On Sprawl

A permanent moratorium on all new major road construction and expansions. Every additional dollar spent building and widening roads digs us deeper into our dangerous oil / auto addiction, and increases global warming.

Trains are by far the most energy efficient form of transportation that greatly reduces global warming, saves lives, and encourages compact, walkable communities.

3. A permanent moratorium on the building of any additional sprawl. Sprawl is probably the single largest contributor to oil addiction and global warming due to it's very design (or lack of). Sprawl forces everyone to drive many miles daily for everything, which in turn requires constant road expansions, encouraging more cars and driving, and more sprawl. Its a vicious cycle consuming ever more oil, and spewing out more pollution, making global warming continually worse.

## New Urbanism

New Urbanism movement promotes the creation and restoration of diverse, walkable, compact, vibrant, mixed-use communities composed of the same components as conventional development, but assembled in a more integrated fashion, in the form of complete communities. These contain housing, work places, shops, entertainment, schools, parks and civic facilities essential to the daily lives of the residents, all within easy walking distance of each other. New Urbanism promotes the increased use of trains and light rail, instead of more highways and roads. Urban living is rapidly becoming the new hip and modern way to live for people of all ages. Currently, there are over 4,000 New Urbanist projects planned or under construction in the United States alone, half of which are in historic urban centers.

It is an international movement to reform the design of the built environment, and is about raising our quality of life and standard of living by creating better places to live. New Urbanism is the revival of our lost art of place-making, and is essentially a re-ordering of the built environment into the form of complete cities, towns, villages, and neighborhoods - the way communities have been built for centuries around the world. New Urbanism involves fixing and infilling cities, as well as the creation of compact new towns and villages.

## The Principles Of New Urbanism

The principles of New Urbanism can be applied increasingly to projects at the full range of scales from a single building to an entire community.

### 1. Walkability

-Most things within a 10-minute walk of home and work

-Pedestrian friendly street design (buildings close to street; porches, windows & doors; tree-lined streets; on-street parking; hidden parking lots; garages in rear lane; narrow, slow speed streets)

-Pedestrian streets free of cars in special cases

### 2. Connectivity

-Interconnected street grid network disperses traffic & eases walking

-A hierarchy of narrow streets, boulevards, and alleys

-High quality pedestrian network and public realm makes walking pleasurable

### 3. Mixed-Use & Diversity

-A mix of shops, offices, apartments, and homes on site. Mixed-use within neighborhoods, within blocks, and within buildings

-Diversity of people - of ages, income levels, cultures, and races

### 4. Mixed Housing

A range of types, sizes and prices in closer proximity

### 5. Quality Architecture & Urban Design

Emphasis on beauty, aesthetics, human comfort, and

creating a sense of place; Special placement of civic uses and sites within community. Human scale architecture & beautiful surroundings nourish the human spirit.

## 6.Traditional Neighborhood Structure

-Discernible center and edge

-Public space at center

-Importance of quality public realm; public open space designed as civic art

-Transect planning: Highest densities at town center; progressively less dense towards the edge.

## 7. Smart Transportation

-Pedestrian-friendly design that encourages a greater use of bicycles, scooters, and walking as means of daily transportation

## 8. Sustainability

-Minimal environmental impact of development and its operations

-Eco-friendly technologies, respect for ecology and value of natural systems

-Energy efficiency

-Less use of finite fuels

-More local production

### Gita-Nagari - The City of the Bhagavad-gita.

As early as 1948, in an unpublished essay entitled Interpretations of Bhagavad-gita, Srila Prabhupada outlined a vision for a city-sized, self-sufficient community based on the spiritual teachings of the Bhagavad-gita. He called the planned community Gita-Nagari, 'the city of the Bhagavad-gita.'

This concept city features an emphasis on spiritual values and cultivation of nonmaterial sources of happiness, a God-centered cosmology, cow protection and ox power, and a spiritually oriented vegetarian diet and division of society according to the principles of varnashrama.

# Solution -15

## Return To
## Land & Cow Economics

*"In the beginning if there is a plot of land and a cow—your whole economic question is solved. Why you should work so hard day and night? So we have created a civilization simply working hard day and night, and the purpose is sense gratification. That's all. That is prohibited. Make your life simplified. Save your time for Krishna consciousness. That is the program. Don't be implicated with sinful activities. Simple life."*
~Srila Prabhupada
   *(Lecture on Srimad-Bhagavatam 5.5.1 — Los Angeles, January 20, 1969)*

More than 50% of world population depends on agriculture and cows for livelihood even today. This way of life is very conducive for preserving ecology, culture and human values.

We wish to quote Srila Prabhupada, founder of the Hare Krishna movement who strongly advocated this God-centred model of living, based on land and cows.

"nayam deho deha-bhajam nrloke kastan kaman arhate vid-bhujam ye.... Kaman means the necessities of life. You can get your necessities of life very easily. By tilling the field, you get grains. And if there is cow, you get milk. That's all. That is sufficient. But the leaders are making plan, that if they are satisfied with their farming work, little grains and milk, then who will work in the factory? Therefore they are taxing so that you cannot live even simple life—this is the position—even if you desire. The modern leaders will not allow you. They force you to work like dogs and hogs and asses."
*(Lecture on Srimad-Bhagavatam 5.5.1-8 — Stockholm, September 8, 1973)*

"That is the system that in India every man is producing his food grains independently. Now it is stopped. Formerly, all these men, they used to produce their food grain. So they used to work for three months in a year, and they could stock the whole year's eatable food grains. Life was very simple. After all, you require to eat. So this Vedic civilization was that keep some land and keep some cows. Then your whole economic question is solved.

Now, in this country, Geneva, I heard there is... I am tasting the milk, first-class milk. I think the world's best milk. Unless one has got his own cows, one cannot get such nice milk. But I hear also that because there is excess production of milk, they have decided to kill twenty-thousand cows.

Just see how much foolish proposal it is. So for want of God consciousness, this mischievous intelligence can be found. The whole economic question can be solved. If you have got excess, then you can trade, you can send to some place where there is scarcity.

What they are doing? In Australia, in Africa, they have got enough land, but the government... Maybe they have no sufficient men to utilize the land, but they won't allow any outsider to go there who can produce. I have seen in Africa. Very, very large tract of land was lying vacant, nobody is producing any food. They are producing coffee. That is not the local men. The Britishers who have gone there, they are producing coffee, tea, and keeping some cows for slaughtering. This is going on. In Australia also, I have seen."
*(Lecture on Bhagavad-gita 13.35 — Geneva, June 6, 1974)*

"Annad bhavanti bhutani. Simply by dry lecture, how they will feel happy? There must be sufficient food grains so that people may live happily, the animal may live happily. Especially in India you will see. No animal is fatty, either cat, dog or cow. They have no eating. So annad bhavanti bhutani. They must be given sufficient food, annad. Krishna does not say that "You fast and chant Hare Krishna." He does not say. Krishna is not so impractical. He says, "Eat very nicely, keep very nicely, and chant Hare Krishna. Make your life successful." That is Hare Krishna consciousness movement. Krishna consciousness movement is not one-sided. It is all-embracing. Sarve sukhino bhavantu. Krishna consciousness movement wants to see everyone happy. Without being happy, how you can remain peaceful?"
*(Lecture on Bhagavad-gita 4.1 — Bombay, March 21, 1974)*

"So our Krishna consciousness movement is not a sentimental movement. It takes care of all-round social organization. It is not something like religious sentiment. Everything should be taken care of. Therefore we say cow protection. Just like this Western civilization has created so may slaughterhouse for eating purposes. But wherefrom they are getting? From mahi , from the land. If there is no pasturing ground, grazing ground, where from they will get the cows and the bulls? Because there is grass on the land and the cows and bulls eat them, therefore they grow. Then you cut their throat, civilized man, and eat, you rascal civilized man. But you are getting from the mahi , from the land. Without land, you cannot. Similarly, instead of cutting the throat of the cows, you can grow your food. Why you are cutting the throat of the cows? After all, you have to get from the mahi , from the land. So as they are, the animal which you are eating, they are getting their eatables from the land. Why don't you get your eatables from the land? Therefore it is said, sarva-kama-dugha mahi . You can get all the necessities of your life from land. So dugha means produce. You can produce your food. Some land should be producing the foodstuff for the animals, and some land should be used for the production of your foodstuffs, grains, fruits, flowers, and take milk. Why should you kill these innocent animals? You take. You keep them muda, happy, and you get so much milk that it will moist, it will make wet the ground. This is civilization. This is civilization."

*(Lecture on Srimad-Bhagavatam 1.10.4 — London, November 25, 1973)*

"In U.S.A. also, there are so much land vacant. They're not utilizing... Whatever production, they... Sometimes they throw it in the water. And I, I have heard in this Geneva, that there was excess of milk production. Therefore they want to kill twenty-thousand cows to reduce the milk production. This is their brain. Actually, there is no brain. So for brain, they should come to these Sastras. They should take guidance. Produce. Produce, utilize. But they'll not utilize. Rather, the limited number of people... At least in India, all the villagers, they have been drawn in the city for producing bolts and nuts. Now eat bolts and nuts.

Solve your problem like... Produce your food wherever you are there. Till little, little labor, and you will get your whole year's food. And distribute the food to the animal, cow, and eat yourself. The cow will eat the refuse. You take the rice, and the skin you give to the cow. From dahl you take the grain, and the skin you give to the... and fruit, you take the fruit, and the skin you give to the cow, and he will give you milk. So why should you kill it? Milk is the miraculous food; therefore Krishna says krsi go raksha vanijya....[Bg. 18.44]. Give protection to the cow, take milk from it, and eat food grains—your food problem is solved. Where is food problem? Why should you invent such civilization always full of anxieties, running the car here and there, and fight with other nation, and economic development? What is this civilization?"

*(Lecture at World Health Organization -- Geneva, June 6, 1974)*

# Solution -16

## Focus On
## Prevention

*"We have a window of only 10 to 15 years to take the steps we need to avoid crossing catastrophic tipping points"*
*–Tony Blair, Prime Minister, UK (October, 2006)*

## Prevention vs Cure

Michael A. Cremo and Mukunda Dasa Goswami, in their classic book 'Divine Nature', explain the paradox of environment preservation: "A group of parents, concerned about the health of their children, will hold a demonstration to remove a toxic-waste dump in the neighborhood. Some say this local, grass-roots approach is the most effective kind of environmental action. In 1991, Hare Krishna members in Poland led a successful grass-roots movement to halt the opening of an environmentally harmful dolomite quarry. But for every grass-roots group that succeeds, dozens fail to overcome the forces arrayed against them. And even if there's success in some particular effort, it has little impact overall. If protesters stop a toxic-waste dump from being set up in one place, it will be set up in another. And the factories that produce the toxic waste will be kept in business by the protesters themselves, who buy what the factories turn out. Furthermore, intense commitment to limited goals and political attempts to achieve them may blind us to the need for the overall spiritual transformation of society."

They continue, "The bigger environmental action organizations, seeking more influence, stage national and even global events, such as Earth Day. They also lobby local, state, and national governments to adopt policies and regulations meant to help solve environmental problems. Many people question the ultimate usefulness of this approach, which has been called "reform environmentalism." It may do some limited good in controlling and reducing pollution, but it leaves intact the whole polluting apparatus of the worldwide industrial society. Also, gains achieved through lobbying and political action can be reversed by the lobbying and political action of others, especially action appealing to economic self-interest. A better approach is to strive for the overall spiritualization of society. With a deep, spiritual change of heart, a permanent change of goals and values, environmental reform would take place as a by-product, almost automatically."

# Solution -17

## Stop Advertising And  Propaganda
## That Fuel Greed, Lust And Consumption

*"Come on, smoke Kool cigarettes and make your brain cool." How can someone become cool by smoking cigarettes? By smoking fire one can become cool? Still, the advertisements are being presented, and the foolish people who are captivated by them smoke cigarettes to become cool. This is maya.*
*~Srila Prabhupada*
*(Dharma – The Way of Transcendence)*

## Oil vs Greed Paradox

Advertising and propaganda has made car driving a symbol of social status. The bigger is your car, the higher is your social standing. Today simplicity is ludicrous and bizarre.

Edward Bernays was one of the first to attempt to manipulate public opinion using the psychology of the subconscious. He felt this manipulation was necessary in society, which he regarded as irrational and dangerous as a result of the 'herd instinct'. He is regarded as father of modern advertising and public relation.

In the 1920s, working for the American Tobacco Company, he sent a group of young models to march in the New York City parade. He then told the press that a group of women's rights marchers would light "Torches of Freedom". On his signal, the models lit Lucky Strike cigarettes in front of the eager photographers. The New York Times (1 April 1928) printed: "Group of Girls Puff at Cigarettes as a Gesture of 'Freedom'". This helped to break the taboo against women smoking in public.

Bernays used his uncle Sigmund Freud's ideas to help convince the public, among other things, that bacon and eggs was the true all-American breakfast.

Bernays helped the Aluminum Company of America (Alcoa) and other special interest groups to convince the American public that water fluoridation was safe and beneficial to human health. This was achieved by using the American Dental Association in a highly successful media campaign. In the 1930s, his Dixie Cup campaign was designed to convince consumers that only disposable cups were sanitary.

In this way, by propaganda, the masses are urged to consume and devour natural resources and a scarcity is the natural outcome.

We want a clean, healthy environment, yet at the same time too many of us demand a style and standard of life that inevitably result in environmental degradation.

Srila Prabhupada, the founder of Hare Krishna movement once said, "Life is never made comfortable by artificial needs, but by plain living and high thinking."

# Solution -18

## Anti-consumerism

*Well, the common enemy in North America is the Western consumer. The consumer has driven oil up to $50 a barrel so we have to have these wars. I think it's incumbent upon us to.*

*–Dan Aykroyd*

# Anti-consumerism

Consumerism refers to a socio-economic trend of twentieth century which consists of mass and mindless consumption of goods and services influenced by massive advertising and centralized marketing and distribution. Consumerism is a also used to describe the effects of the market economy on the individual.

Consumer economy leads to mass craze for goods and services, and the devaluing of the worth of a good or service, and instead focusing on its price in the market. In many critical contexts the term is used to describe the tendency of people to identify strongly with products or services they consume, especially those with commercial brand names and obvious status-enhancing appeal, e.g. an expensive automobile or jewelry. A culture that has a high amount of consumerism is referred to as a consumer culture.

Over-consumption is threatening emotional destabilization of the global population. To those who embrace the idea of consumerism, these products are not seen as valuable in themselves, but rather as social signals that allow them to aggrandize themselves through consumption and display of similar products.

Anti-consumerism refers to the sociopolitical movement against consumerism. At the heart of consumerism lie large business corporations who are invading people's privacy, manipulating politics and governments and creating false needs in consumers. Their modus operandi includes invasive advertising (adware, spam, telemarketing, etc.), massive corporate campaigns, contributions in democratic elections, interference in the policies of sovereign nation states, and endless global news stories about corporate corruption.

Concern over the treatment of consumers has led to widespread movement to educate consumers and protect consumers' rights. Anti-consumerist activism draws some parallels with environmental activism, anti-globalization, and animal-rights activism in its condemnation of modern corporate organizations. In recent years, there have been an increasing number of books and films, which have to some extent popularized an anti-corporate ideology in the public.

Opposition to economic materialism comes primarily from two sources: religion and social activism. Some religions assert that materialism interferes with the connection between the individual and the divine, or that it is inherently an immoral lifestyle. Some notable individuals, such as Francis of Assisi, Ammon Hennacy, and Mohandas Gandhi, have claimed that spiritual inspiration led them to a simple lifestyle. Social activists have asserted that materialism is connected to war, crime, and general social malaise. Fundamentally, their concern is that materialism is unable to offer a proper raison d'être for human existence.

Many anti-corporate activists believe that the rise of large business corporations is posing a threat to the legitimate authority of nation states and the public sphere. Malpractices by corporations is a norm rather than an exception. Corporations' responsibility is to answer only to shareholders, giving human rights and other issues almost no consideration. Multinational corporations will usually pursue strategies that intensify labor and attempt to reduce costs. For example, they will (either directly, or through subcontractors) attempt to find low wage economies with laws which are conveniently lenient on human rights, the natural environment, trade union organization and so on.

Consumer economies are governed not by production but by consumption, and that the advertising techniques used to create consumer behavior amount to the destruction of psychic and collective individuation, an addictive cycle of consumption, leading to hyper-consumption, the exhaustion of desire, and the en masse prevalence of discontent and dissatisfaction.

The term and concept of conspicuous consumption originated at the turn of the 20th century in the writing of economist Thorstein Veblen. The term describes an apparently irrational and confounding form of economic behaviour. Veblen's scathing proposal is that this unnecessary consumption is a form of status display.

Overcoming Consumerism is a growing philosophy. It is a term that embodies the active resistance to consumerism. It is being used by many universities as a term for course material and as an introduction to the study of marketing from a non-traditional approach.

Consumerism impacts far beyond the immediate consumer group. There is obvious link between the relentless consumerism advocated by both governments and advertisers, and the continued degradation and destruction of the natural environment and fuel shortages.

# Solution -19

## New Age Movements And
## Under-Currents

*The stars, that nature hung in heaven, and filled their lamps with everlasting oil, give due light to the misled and lonely traveller.*
*–John Milton*

## Current Social Trends

The Trends Research Institute of Rhineback, New York has determined that simplifying lifestyles is one of the leading movements of the 1990s. They estimate that by the end of the decade, 15% of the 77 million American will have made significant movements towards living simpler lives. A 1995 survey showed that about 30% of Americans had downshifted voluntarily, many working fewer hours for less pay so that they could spend more time with the family. Other surveys have shown that 60 to 80% of workers would be willing to accept reductions in pay if they could work fewer hours.

In 1995, the Merck Family Fund commissioned a major study into the issues of consumption in the U.S. It showed that when people were asked to describe what they were looking for in life, their aspirations rarely centred on material goods. The things they really wanted were nonmaterial. Topping the list, 66% of the people surveyed said they would be much more satisfied with their lives "if I were able to spend more time with my family and friends". 55% said they would be more satisfied "if there was less stress in my life", and 47% said "if I felt like I was doing more to make a difference in my community". Just 21% answered by saying "if I had a nicer car", 19% by saying "if I had a bigger house or apartment", and only 15% by saying "if I had more nice things in my home".

Following are some of the trends that have emerged in last few decades as an answer to onslaught on ecology by the industrial civilization. All these ideas mentioned below will help relieve global oil crunch.

### Ecovillage

Ecovillage refers to a small community or habitation which is environment friendly, often consisting of 50-150 individuals. In this form of full-featured settlement, human activities are harmlessly integrated into the natural world that is supportive of healthy human development, and which can be successfully continued into the indefinite future.

The habitants of an ecovillage often share identical ecological, social or spiritual values and they have chosen an alternative to a lifstyle featuring wasteful consumerism, the destruction of natural habitat, urban sprawl, factory farming, and over-reliance on fossil fuels. Averting ecological disaster is at the forefront of this concept.

Larger ecovillages of up to 2,000 individuals may, however, exist as networks of smaller "ecomunicipalities" or subcommunities to create an ecovillage model that allows for social networks within a broader foundation of support.

Rural ecovillages are usually based on organic farming, permaculture and other approaches which promote ecosystem function and biodiversity. Some ecovillages integrate many of the design principles of cohousing, but with a greater ecological focus and a more organic  process, typical of permaculture design.

A directory of ecovillages is appended at the end.

## Back-To-The-Land Movement

Back-to-the-land movement refers to migration from cities to rural areas. This was a social phenomenon of the 1960s and 1970s in North America. Its roots are in fact European and can be traced back to the Romantics and beyond.

There have been back-to-the-land population movements down through the centuries. These have happened in different parts of the world, largely due to the occurrence of severe urban problems and people's felt need to live a better life, often simply to survive. For example with the fall of Rome, city dwellers re-inhabited the rural areas of the region.

In recent past, economic theorist like Ralph Borsodi is said to have influenced thousands of urban-living people to try a modern homesteading life during the Great Depression.

There was again a fair degree of interest in moving to rural land after World War II. In 1947 Betty MacDonald published what became a popular book, The Egg and I, telling her story of marrying and then moving to a small farm on the Olympic Peninsula in Washington state.

After World War II, many returned veterans sought a meaningful life far from the ignobility of modern warfare and industrial city life, and moved to semi-wilderness environs.

But moving back to land was a special phenomenon of the '60s and '70s. The movement was sparked off by a book named 'Living the Good Life' by Scott Nearing.

By the late '60s, many people were attracted to getting more in touch with the basics of life (for instance, what a potato plant looks like, or the act of milking a cow) — after they felt out of touch with nature, in general. A distinct social trend was visible which reflected people's fatigued attitude towards rampant consumerism, the failings of government and society, urban deterioration, air and water pollution, struggle or boredom of "moving up the company ladder." Paralleling the desire for reconnection with nature was a desire to reconnect with physical work, for many were drawn by some sense of dignity in physical labor, just as they might feel depressed contemplating the prospect of a worklife at a desk in the city.

Counterculture of the 1960s fuelled back-to-land movement to some degree. Most of the back-to-the-landers wanted greater contact with nature, and sought to become self-employed workers in a cottage industry. Many wished to build their own house, and produce a good deal of their own food. 'Knowing your neighbor' was an important aspect of the movement in contrast with isolated city life. Another common practice to surface was the incorporation of barter, a form of trade where goods or services from one individual or household are exchanged for a certain amount of other goods or services from another individual or household, and in which no money is involved in the transaction.

Organic horticulture and organic agriculture are integral aspects of the back-to-the-land movement. All over world organics is a fad today. American consumers spent $1 billion on organically grown food in 1994, and $13 billion in 2003.

In the 1990s the term "urban refugees" became popular in relation to back-to-the-landers.

## Environmental Movement

The environmental movement advocates the sustainable management of resources and stewardship of the natural environment through changes in public policy and individual behavior. This movement works from diverse platforms like scientific, social, and political. The movement is centered around ecology, health, and human rights.

The environmental movement is represented by a range of organizations, from the large to grassroots. It has a large membership of varying and strong beliefs which include private individuals, professionals, religious leaders, politicians, and extremists.

The roots of the modern environmental movement can be traced to attempts in nineteenth-century Europe and North America to expose the costs of environmental negligence, notably disease, as well as widespread air and water pollution, but only after the Second World War did a wider awareness begin to emerge.

In the United States two early conservationists stood out as leaders in the movement; Henry David Thoreau and George Perkins Marsh. Thoreau was concerned about the wildlife from Massachusetts. He wrote Walden as he studied the wildlife from a cabin. Marsh was influential with regards to the need for resource conservation.

During the 1950s, 1960s, and 1970s, several events illustrated the magnitude of environmental damage caused by man. In 1954, the 23 man crew of the Japanese fishing vessel Lucky Dragon was exposed to radioactive fallout from a hydrogen bomb test at Bikini Atoll. In 1962 the publication of the book Silent Spring by Rachel Carson drew attention to the impact of chemicals on the natural environment. In 1967 the Torrey Canyon oil tanker went aground off the southwest coast of England, and in 1969 oil spilled from an offshore well in California's Santa Barbara Channel. In 1971 the conclusion of a law suit in Japan drew international attention to the effects of decades of mercury poisoning on the people of Minamata.

In 1972, the United Nations Conference on the Human Environment was held in Stockholm, and for the first time united the representatives of multiple governments in discussion relating to the state of the global environment. This conference led directly the creation of government environment agencies and the UN Environment Program. The United States also passed new legislation such as the Clean Water Act, the Clean Air Act, the Endangered Species Act, and the National Environmental Policy Act- the foundations for current environmental standards.

Since the 1970s, public awareness, environmental sciences, ecology, and technology have advanced to include modern focus points like ozone depletion, global climate change, acid rain, and the harmful potential of genetically modified organisms (GMOs).

## Conservation Movement

The conservation movement as a concept aims to preserve natural resources expressly for their continued sustainable use by humans. Its a political, social and, to some extent, scientific movement that seeks to protect natural resources including plant and animal species as well as their habitat for the future.

The early conservation movement included fisheries and wildlife management, water, soil conservation and sustainable forestry. The contemporary conservation movement has broadened from the early movement's emphasis on use of sustainable yield of natural resources and preservation of wilderness areas to include preservation of biodiversity. In other parts of the world conservation is used more broadly to include the setting aside of natural areas and the active protection of wildlife for their inherent value, as much as for any value they may have for humans.

## Permaculture

Permaculture refers to sustainable and earth friendly agricultural systems, which are distinct from destructive industrial-agricultural methods that poison the land and water, reduce biodiversity and remove billions of tons of soil from previously fertile landscapes. The term permaculture basically means "permanent agriculture" but this can also be taken to mean "permanent culture".

This concept was presented during the 1970s by two Australians, Bill Mollison and David Holmgren. Through a series of publications, Mollison, Holmgren and their associates documented an approach to designing human settlements, in particular the development of perennial agricultural systems that mimic the structure and interrelationship found in natural ecologies.

Permaculture aims to promote self-sufficient human settlements — ones that reduce society's reliance on industrial systems of production and distribution that fundamentally and systematically are destroying the earth's ecosystems.

By the early 1980s, the concept had moved on from being predominantly about the design of agricultural systems towards being a more fully holistic design process for creating sustainable human habitats.

By the mid 1980s, many of the students had become successful practitioners and had themselves begun teaching the techniques they had learned. In a short period of time permaculture groups, projects, associations, and institutes were established in over one hundred countries.

There are now two strands of permaculture: a) Original and b) Design permaculture. Original permaculture attempts to closely replicate nature by developing edible ecosystems which closely resemble their wild counterparts. Design permaculture takes the working connections at use in an ecosystem and uses them as its basis. The end result may not look as "natural" as a forest garden, but still has an underlying design based on ecological principles.

### Green Movement

The Green movement refers to political forums in various countries which aim at protecting environment and promoting sustainability, social justice, ecology, conservation, peace and nonviolence. These groups are called greens.

This movement started in March of 1972, with the world's first green party, the United Tasmania Group in Australia. First Greens to contest and win elections were the German Greens, in 1980. The German Greens drew support for their opposition to nuclear power, pollution, and the actions of NATO.

In Finland, in 1995, the Green League became the first European Green party to form part of a state-level Cabinet. The German Greens followed, forming a government with the Social Democratic Party of Germany (the "Red-Green Alliance") from 1998 to 2005. In 2001, they reached an agreement to end reliance on nuclear power in Germany.

There is a growing level of global cooperation between Green parties. Global gatherings of Green Parties now happen. The first Planetary Meeting of Greens was held in Rio de Janeiro, immediately preceding the United Nations Conference on Environment and Development held there. More than 200 Greens from 28 nations attended. The next Global Green Gathering was held in Nairobi, Kenya in 2008

In 1996, 69 Green Parties from around the world signed a common declaration opposing French nuclear testing in the South Pacific, the first statement of global greens on a current issue. A second statement was issued in December 1997, concerning the Kyoto climate change treaty.

Separately from the Global Green Gatherings, Global Green Meetings take place. For instance, one took place on the fringe of the World Summit on Sustainable Development in Johannesberg. Green Parties attended from Australia, Taiwan, Korea, South Africa, Mauritius, Uganda, Cameroon, Republic of Cyprus, Italy, France, Belgium, Germany, Finland, Sweden, Norway, the USA, Mexico and Chile.

The member parties of the Global Greens are organised into four continental federations:

Federation of Green Parties of Africa

Federation of the Green Parties of the Americas / Federación de los Partidos Verdes de las Américas

Asia-Pacific Green Network

European Federation of Green Parties

## Neo-Tribalism

Neo-Tribalism philosophy propounds that tribal community setup in close proximity with nature is a natural and more conducing way of life for human beings. It also asserts that all problems faced by human society are byproducts of modern living and it cannot achieve genuine happiness until some semblance of tribal lifestyles has been re-created or re-embraced.

Neo-tribalist ideology is rooted in the social philosophies of Jean-Jacques Rousseau and William Kingdon Clifford, who spoke of a "tribal self" thwarted by modern society. A species removed from the environment in which it evolved, in which it is meant to live, will become pathological, has been cited by Neo-tribalists as providing a scientific basis for their beliefs.

Certain aspects of industrial and post-industrial life, including the necessity of living in a society of strangers and interacting with organizations that have very large memberships, are cited as inherently detrimental to the human mind. In a 1985 paper, "Psychology, Ideology, Utopia, & the Commons," psychologist Dennis Fox proposed a number around 150 people for an ideal settlement.

Advocates attribute a general breakdown in the social structure of modern civilization to more frequent moves for economic reasons, longer commutes and a lack of strong friendships and community bonds.

The French Sociologist Michel Maffesoli was perhaps the first to use the term neo-Tribalism in a scholarly context. Maffesoli predicted that as the culture and institutions of modernism declined, societies would look to the organizational principles of the distant past for guidance, and that therefore the post-modern era would be the era of Neo-Tribalism.

Radical neo-Tribalists such as John Zerzan believe that healthy tribal life can only thrive after technological civilization has either been destroyed or severely reduced in scope. Daniel Quinn, associated with the New tribalists, formulated the concept of "walking away": abandoning the owner/conqueror worldview of civilization - though not necessarily its geographical space - and making a living with others in tribal businesses. Others, such as Derrick Jensen, tend to call for more violent action, as they believe that it is appropriate and necessary to actively accelerate or cause a collapse of civilization. Still others, such as The Tribe of Anthropik take a survivalist bent and believe that a collapse is inevitable no matter what is done or said and concentrate their efforts on surviving and forming tribal cultures in the aftermath.

## Urban Tribalism

This is a moderate form of neo-Tribalism. Its followers believe that a tribal social structure can coexist with a modern technological infrastructure. For example, under this scenario, people might reside in a large house or other building with a communal group of 12-20 individuals all abiding by a defined set of rules, cultural rituals and intimate relationships, but otherwise leading modern lives, going to a job, driving a car, etc. In that it attempts to harmonize two seemingly contradictory cultures, namely modern existence and tribalism.

## Drop City

This was a short-lived movement in 60s but before fading away it inspired several moves in the field of alternative living. Its residents were called 'droppers'.

Drop City was an artists' community that formed in southern Colorado in 1965.

It attracted people from around the world who came to stay and work on the construction projects. Inspired by the architectural ideas of Buckminster Fuller, residents constructed dome like structure to house themselves, using geometric panels made from the metal of automobile roofs and other inexpensive materials. Soon the community grew in reputation and size, accelerated by media attention, including news reports on national television networks.

## Intentional Community

An intentional community is a planned residential community designed to promote voluntary simplicity, interpersonal growth, self-reliance, sharing of resources, creating family-oriented neighborhoods and living ecologically sustainable lifestyles. Its distinctive feature is a much higher degree of social interaction than other communities. The members of an intentional community typically hold a common social, political or spiritual vision. They also share responsibilities and resources. Intentional communities include cohousing, residential land trusts, ecovillages, communes, kibbutzim, ashrams, and housing cooperatives. Some communities are secular; others have a spiritual basis.

A survey in the 1995 edition of the Communities Directory, published by Fellowship for Intentional Community (FIC), reported that 54% of the communities choosing to list themselves were rural where as 28% were urban.

### Buy Nothing Day : Shop Less - Live More

Buy Nothing Day (BND) is an international day of protest against consumerism observed by social activists. Typically celebrated the Friday after American Thanksgiving in North America and the following day internationally. It was founded in Vancouver by artist Ted Dave and subsequently promoted by Adbusters magazine, based in Canada.

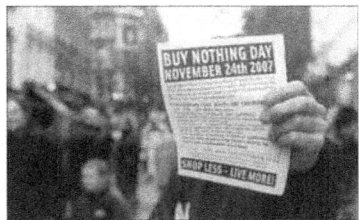

The first Buy Nothing Day was organized in Mexico in September 1992 "as a day for society to examine the issue of over-consumption." In 1997, it was moved to the Friday after American Thanksgiving, also called "Black Friday", which is one of the ten busiest shopping days in the United States. Outside North America and Israel, Buy Nothing Day is the following Saturday. Adbusters was denied advertising time by almost all major television networks except for CNN, which was the only one to air their ads. Soon, campaigns started appearing in the United States, the United Kingdom, Israel, Austria, Germany, New Zealand, Japan, the Netherlands, France, and Norway. Participation now includes more than 65 nations.

It's a day where you challenge yourself, your family and friends to switch off from shopping and tune into life. The rules are simple, for

24 hours you will detox from consumerism and live without shopping. Anyone can take part provided they spend a day without spending!

**Activities**

Various gatherings and forms of protest have been used on Buy Nothing Day to draw attention to the problem of over-consumption:

-*Credit card cut up*: Participants stand in a shopping mall, shopping center, or store with a pair of scissors and a poster that advertises help for people who want to put an end to mounting debt and extortionate interest rates with one simple cut.

-*Zombie Walk*: Participant "zombies" wander around shopping malls or other consumer havens with a blank stare. When asked what they are doing participants describe Buy Nothing Day.

-*Whirl-mart*: Participants silently steer their shopping carts around a shopping mall or store in a long, baffling conga line without putting anything in the carts or actually making any purchases.

-*Wildcat General Strike*: A strategy used for the 2009 Buy Nothing Day where participants not only do not buy anything for twenty-four hours but also keep their lights, televisions, computers and other non-essential appliances turned off, their cars parked, and their phones turned off or unplugged from sunrise to sunset.

-*Buy Nothing Day hike*: Rather than celebrating consumerism by shopping, participants celebrate The Earth and nature.

### National Downshifting Week

The 'grassroots' awareness campaign, National Downshifting Week (UK) (founded 1995) encourages participants to positively embrace living with less. Campaign creator, British writer and broadcaster on downshifting and sustainable living, Tracey Smith says, "The more money you spend, the more time you have to be out there earning it and the less time you have to spend with the ones you love". National Downshifting Week encourages participants to 'Slow Down and Green Up' and contains a list of suggestions for Individuals, Companies and Children and Schools to help them lean towards the green, develop corporate social responsibility in the workplace and create eco-protocols and policies that work alongside the national curriculum, respectively.

## Guerrilla Gardening

Guerrilla gardening is political gardening, a form of nonviolent direct action, primarily practiced by environmentalists. Activists take over an abandoned piece of land which they don't own to grow crops or plants. The practices are non-violent, unlike guerrilla warfare that can cause bloodshed. Guerrilla gardeners believe in reclaiming land from perceived neglect or misuse and assigning a new purpose for it.

Guerrilla gardeners will sometimes carry out their actions late at night geared up with gardening gloves, watering cans, compost, seeds and plants. They plant and sow a new vegetable patch or flowering garden. Others will work more openly, actively seeking to engage with members of the local community.

There is a community where you can advertise your planned attack, here http://guerrillagardening.org/community/index.php

## World Brotherhood Colonies

World Brotherhood Colonies is an idea for cooperative spiritual living, first promoted by Paramahansa Yogananda as early as in 1932. Yogananda urged young people to pool their resources, buy land and build spiritual communities where they could live a life of "plain living and high thinking." Yogananda tried to establish World Brotherhood Colonies at his retreat centers in Southern California. They eventually were turned into monasteries for monks and nuns of Self-Realization Fellowship.

Yogananda often emphasized the need for intentional communities "founded on a spiritual basis." His vision for Colonies included couples, families, and single people sharing a cooperative community life, with the common bond of daily meditation and selfless service. He felt that Colonies would have a far-reaching effect on modern society:

"Man is a soul, not an institution; his inner reforms alone can lend permanence to outer ones. By stress on spiritual values, self-realization, a colony exemplifying world brotherhood is empowered to send inspiring vibrations far beyond its locale."

A unique feature of Yogananda's World Brotherhood Colonies idea was that it offered married people and families a spiritually fulfilling community life. Many traditional monastic communities and ashrams offer most of the features of Yogananda's Colonies - simple living, selfless service, cooperation, and daily meditation.

Master envisioned the idea as one in which all may work together in a self-supporting group wherein each one is dedicated to God.

Yogananda often spoke of the practical benefits that come from cooperative living. Even though he was a teacher of meditation and yoga, he frequently gave practical advice on subjects such as diet, exercise, business, education, and prosperity. As early as 1932, he urged his students to avoid buying cars and other luxuries on the 'installment plan', similar to the modern credit card. As per him, living in such communities would help people be free of many of the ills that beset modern society.

He also wrote, in an early correspondence course:

"Gather together, those of you who share high ideals, pool your resources. Buy land out in the country. A simple life will bring you inner freedom. Harmony with nature will bring you a happiness known to few city dwellers. In the company of other truth seekers it will be easier for you to meditate and think of God.

What is the need for all the luxuries people surround themselves with? Most of what they have they are paying for on the installment plan. Their debts are a source of unending worry to them. Even people whose luxuries have been paid for are not free; attachment makes them slaves. They consider themselves freer for their possessions, and don't see how their possessions in turn possess them!

Let every man gather from five to ten thousand dollars, and, in groups of thirty, let them build self-sustaining, self-governing colonies, starting with California. Do not spend the principal of the money, except what is necessary to buy land and to start the colony. Put the money in a trust fund. Pay taxes with the interest. If taxes were abolished, people could live by exchange.... Time should not be wasted in producing luxuries.

Start now building colonies, and stop industrially selfish society from gambling with your destiny. Get away from the perpetual slavery of holding jobs to the last day of your life. Buy farms and settle down with harmonious friends. Work three hours a day and live in the luxury of literary wealth, and have time to constructively exchange Divine experiences and meditate."

In that article, he gave five guidelines for people living in World Brotherhood Colonies:

Cut down luxuries.

Think yourself a child of God.

Think of all nationalities as your brothers.

Seek prosperity for yourself and for others.

Develop the creative thought of success every day after deep meditation.

Yogananda was so enthusiastic about the idea, that he once said, "I was thinking so much last night about world brotherhood colonies that my mind didn't want to meditate. Then I chanted a little bit, and my mind came back to me." He also wrote a letter to Henry Ford, founder of the Ford Motor Company, trying to elicit Ford's support for World Brotherhood Colonies. He felt so strongly about the idea, that he once declared, "The day will come when this colony idea will spread through the world like wildfire."

Swami Kriyananda wrote that a talk given by Yogananda on World Brotherhood Colonies was the most stirring lecture he ever heard. The occasion was a garden party in Beverly Hills, in July 1949:

"This day," he (Yogananda) thundered, punctuating every word, "marks the birth of a new era. My spoken words are registered in the ether, in the Spirit of God, and they shall move the West.... Self-Realization has come to unite all religions.... We must go on — not only those who are here, but thousands of youths must go North, South, East and West to cover the earth with little colonies, demonstrating that simplicity of living plus high thinking lead to the greatest happiness!"

On August 20, 1950, Yogananda dedicated the Self-Realization Fellowship Lake Shrine and Mahatma Gandhi World Peace Shrine at Pacific Palisades, California. He spoke to over 1500 people gathered there for the event. Much of his lecture revolved around what he called "the art of living", which included advice on how to find spiritual happiness, and how to achieve harmony between all people and religions. He spoke about World Brotherhood Colonies as an important aspect of the art of living:

"There must be world brotherhood if we are to be able to practice the true art of living, and in this connection I wish to emphasize four points."

"...we must build colonies wherein we can take youths who are 100% willing and give them character education and the opportunity to find happiness, freedom, job, home, and church all in one place, and to produce food for their own use. We have started trial colonies at Encinitas and at Mt. Washington in Los Angeles, and we have some colonies in India too. Ours is not a church in the ordinary sense. We never ask anyone if he is Jew, Gentile, Mohammedan, or Catholic — willingness and good character are the criterion of acceptance...."

"The colony system is succeeding because it isn't simply church for an hour, but working together all the time for mutual good. We are not making empty members. All the ministers and many of the members have helped in the building of this SRF Lake Shrine. That is the secret. We must work for God and commune with Him. I believe that America is a wonderful country in which to try out brotherhood colonies, wherein mankind can learn that the first principle of life is happiness."

In his autobiography, he wrote, "He hath made of one blood all nations of men. An urgent need on this war-torn earth is the founding, on a spiritual basis, of numerous world-brotherhood colonies. 'World brotherhood' is a large term, but man must enlarge his sympathies, considering himself in the light of a world citizen. He who truly understands that "it is my America, my India, my Philippines, my Europe, my Africa" and so on, will never lack scope for a useful and happy life."

In 1968, Swami Kriyananda, a disciple of Yogananda, started the first Ananda community outside Nevada City, California, based on Yogananda's World Brotherhood Colonies principles. As of 2007, Ananda Village has grown to 840 acres, with 250 residents. The community includes schools (kindergarten through high school), private and community-owned businesses, gardens, a guest retreat and teaching center, a healing center, a museum and gift shop, publishing company, and more. Since the founding of Ananda Village, Ananda has begun six more World Brotherhood Colonies. As of 2007, there were approximately 1,000 residents living in the Ananda World Brotherhood Colonies.

## Anarcho-Primitivism

Anarcho-primitivists advocate a return to non-"civilized" ways of life through deindustrialisation, abolition of division of labour or specialization, and abandonment of technology. Anarcho-primitivists are often distinguished by their focus on achieving a feral state through "rewilding".

Within the last half-century, societies once viewed as barbaric have been largely reevaluated by academics, some of whom now hold that early humans lived in relative peace and prosperity. Frank Hole, an early-agriculture specialist, and Kent Flannery, a specialist in Mesoamerican civilization, have noted that, "No group on earth has more leisure time than hunters and gatherers, who spend it primarily on games, conversation and relaxing."

## The Affluenza Theory

### Victory of Perceived Wants Over Real Needs

Affluenza refers to an epidemic of stress, overwork, waste and indebtedness caused by the pursuit of the American Dream and an unsustainable addiction to economic growth. It is a social pathologic condition caused by overload, debt, anxiety and waste resulting from the dogged pursuit of more. Thus affluenza is placing an unreasonably high value on money, possessions, appearances (physical and social) and fame.

This theory proposes that the costs of acquiring material wealth vastly outweigh the benefits. Those who become wealthy will find the economic success leaving them unfulfilled and hungry for more wealth.

This insatiable greed for more and more possessions is leading to social inequality, stimulation of artificial needs, overconsumption, "luxury fever", consumer debt, overwork, waste, and harm to the environment. These pressures lead to "psychological disorders, alienation and distress", causing people to "self-medicate with mood-altering drugs and excessive alcohol consumption"

Higher rates of mental disorders are the consequence of excessive wealth-seeking in consumerist nations. As per World Health Organization, English-speaking nations have twice as much mental illness as mainland Europe: 23% vs 11.5% due to lesser prevalence of affluenza in mainland Europe as compared to English-speaking nations.

Affluenza is considered to be most present in the United States, where the culture encourages its citizens to measure their worth by financial success and material possessions. Mainstream media outlets, such as television, tend to demonstrate how pervasive the idea has become; and by the same token, the same media outlets reinforce the values to the viewers. Affluenza also tends to bring very high social costs and environmental strain by diminishing endangered natural resources like oil.

Proof of this disease lies in the fact that the economy is doing so well, but we are we not becoming happier. The solution lies when people start defining themselves as having value independent of their material possessions.

## Freeganism

The word "freegan", is coined out of "free" and "vegan". Freeganism started in the mid 1990s, out of the antiglobalization and environmentalist movements.

Freeganism is an anti-consumerism lifestyle whereby people employ alternative living strategies based on "limited participation in the conventional economy and minimal consumption of resources. Freegans embrace community, generosity, social concern, freedom, cooperation, and sharing in opposition to a society based on materialism, moral apathy, competition, conformity, and greed." The lifestyle involves salvaging discarded, unspoiled food from supermarket dumpsters that have passed their sell by date, but are still edible and nutritious. They salvage the food not because they are poor or homeless, but as a political statement.

Freegans' motivations are varied and numerous; some adhere to it for environmental reasons, some for religious reasons and others do it to embrace the philosophy as a form of political consciousness.

Many freegans get free food by pulling it out of the trash, a practice commonly nicknamed 'dumpster diving' in North America and 'skipping or bin diving' in the UK. Freegans find food in the garbage of restaurants, grocery stores, and other food-related industries, and this allows them to avoid spending money on products that they claim exploit the world's resources, contribute to urban sprawl, treat workers unfairly, or disregard animal rights. By foraging, they believe that they are preventing edible food from adding to landfills and sometimes feed people and animals who might otherwise go hungry.

Instead of buying industrially grown foods, wild foragers find and harvest food and medicinal plants growing in their own communities. Some freegans participate in "Guerrilla" or "Community" Gardens, with the stated aim of rebuilding community and reclaiming the capacity to grow one's own food. They claim to seek an alternative to dependence and participation in what they perceive as an exploitative and ecologically destructive system of global, industrialized corporate food production.

Sharing is also a common freegan practice. "Food Not Bombs" recovers food that would otherwise go to waste to serve warm meals on the street to anyone who wants them. The group promotes an ethic of sharing and community while working to show what they consider to be the injustice of a society in which fighting wars is considered a higher priority than feeding the hungry.

"Really, Really Free Markets" are free social events in which freegans can share goods instead of discarding them, share skills, give presents and eat food.

Freegans also advocate sharing travel resources. Internet-based ridesharing reduces the use of cars and all the related resources needed to maintain and operate them.

## Community Bicycle Programs

This is an international movement to promote environmentally friendly transportation. The central concept is free (or nearly free) access to bicycles for inner-city transport. These programs appear in all shapes and sizes in cities throughout the world. The goal is to reduce the use of automobiles for short trips inside the city and diminish traffic congestion, noise and air-pollution.

Community Bike Programs and Bike Collectives facilitate community sharing of bicycles, restore found and broken bikes, and teach people how to do their own bike repairs. In the process they build a culture of skill and resource sharing, reuse wasted bikes and bike parts, and create greater access to environmentally friendly transportation.

These programs are also known as Yellow bicycle programs, White bicycle programs, bike sharing, public bike or free bike.

The bikes can be returned at any station in the system, which facilitates one way rides to work, education or shopping centres. Thus, one bike may take 10-15 rides a day with different users and can be ridden up to 10,000 km (6000 miles) a year.

## Frugality

Frugality in the context of certain belief systems, is a philosophy in which one does not trust, or is deeply wary of "expert" knowledge, often from commercial markets or corporate cultures, claiming to know what is in the best economic, material, or spiritual interests of the individual.

There are many different spiritual communities that consider frugality a virtue or a spiritual discipline. The Religious Society of Friends and the Puritans are examples of such groups. The basic philosophy behind this is the idea that people ought to save money in order to allocate it to more charitable purposes, such as helping others in need.

There are also environmentalists who consider frugality to be a virtue through which humans can make use of their ancestral skills as hunter-gatherers, carrying little and needing little, and finding meaning in nature instead of man-made conventions or religion. Henry David Thoreau expressed a similar philosophy in Walden, with his zest for self-reliance and minimal possessions while simply living in the woods.

## Sustainable Living

Sustainable living might be defined as a lifestyle that could be sustained without exhausting natural resources. The term can be applied to individuals or societies. This concept is generally applied in the areas of transport, housing, energy, and diet. Sustainable living aims at shifting from a fossil fuel-based, automobile-centered economy to a renewable energy-based, diversified transport and reuse/recycle economy.

Sustainable living is a sub-division of sustainability where the prerequisites of a modern, industrialized society are left unexercised by choice for a variety of reasons.

The publication of Living the Good Life by Helen and Scott Nearing in 1954 is the modern-day beginning of the sustainability movement. The book fostered the back-to-the-land movement in the late 1960s and early 1970s.

In 1972, Donella Meadows wrote the international bestselling book 'The Limits to Growth', which reported on a study of long-term global trends in population, economics, and the environment. It sold millions of copies and was translated into 28 languages.

### Plain Dress

Plain dress is a religious practice, developed in the late 18th century, where people dress in clothes of simple design, simple fabric, and unrevealing cut. It is used to indicate the simplicity of the wearers life, and the separateness from the rest of the world. It is practiced by some Amish, Apostolics, Brethren, Mennonites, and Friends (Quakers).

### Cohousing

A cohousing community is planned, owned and managed by the residents, groups of people who want more interaction with their neighbours and avail economic and environmental benefits of sharing resources, space and items.

A cohousing community is a kind of intentional community composed of private homes with basic facilities supplemented by extensive common facilities. Common facilities vary but usually include a large kitchen and dining room where residents can take turns cooking for the community. Other facilities may include a laundry, pool, child care facilities, offices, internet access, game room, TV room, tool room or a gym. Through spatial design and shared social and management activities, cohousing facilitates intergenerational interaction among neighbours, for the social and practical benefits.

The modern theory of cohousing originated in Denmark in the 1960s among groups of families who were dissatisfied with existing housing and communities that they felt did not meet their needs. Bodil Graae published "Children Should Have One Hundred Parents," spurring a group of 50 families to organize around a community project in 1967. This group developed the cohousing project Sættedammen, which is the oldest known cohousing community in the world. The first community in the United States to be designed, constructed and occupied specifically for cohousing is Muir Commons in Davis, California.

Hundreds of cohousing communities exist in Denmark and other countries in northern Europe. There are nearly 100 operating communities in the United States with more than 100 others in the planning phases. In Canada, there are 7 completed communities, and approximately 15 in the planning/construction process. There are also communities in Australia, the UK and other parts of the world.

This form of living goes a long way in conserving the resources and protecting the environment. This form of habitation allows more space for social forestry.

## Survivalism

Survivalism is an individual endeavour to survive a collapse of national system either due to economic failure or nuclear exchange etc. A survivalist prepares to lay low during a socio-economic collapse by migrating to low risk areas and building up inventory of essentials and if required fortify the hideout for possible pillage.

In the 60's, this concept rose to prominence with concerns of monetary devaluation, possible nuclear exchanges between the US and the Soviet Union and vulnerability of urban centers.

In 1976, survival bookseller and author Don Stephens in Washington (author of The Survivor's Primer & Updated Retreater's Bibliography, 1976) popularized the term "retreater" and advocated relocating to a rural retreat when society breaks down.

Survivalist retreat books of the 1980s were typified by the 1980 book, 'Life After Doomsday' by Bruce D. Clayton advocating survival retreats in locales that would minimize fallout and specially-constructing Blast shelters and/or Fallout shelters that would provide Fallout Protection in the event of a nuclear war.

In recent years, advocates of survivalist retreats have had a strong resurgence after the terrorist attacks on the World Trade Center in New York in 2001 and similar attacks in Bali, Spain, and London.

Online survival websites, forums, and blogs discuss the best locales for survival retreats, how to build, fortify, and equip them, and how to form survivalist retreat groups. In his books and in his blog (SurvivalBlog), James Wesley Rawles uses the generic term "beans, bullets and Band Aids" to describe retreat logistics.

## Social Ecology

Social ecology proposes that at the root of all environmental ills lie an overly competitive grow-or-die philosophy which is rooted in dominatory hierarchical political and social systems. It suggests that this cannot be resisted by individual action such as ethical consumerism but must be addressed by more nuanced ethical thinking and collective activity grounded in radical democratic ideals. The complexity of relationships between people and with nature is emphasised, along with the importance of establishing social structures that take account of this.

What literally defines social ecology as "social" is its recognition of the often overlooked fact that nearly all our present ecological problems arise from deep-seated social problems. Conversely, present ecological problems cannot be clearly understood, much less resolved, without resolutely dealing with problems within society. To make this point more concrete: economic, ethnic, cultural, and gender conflicts, among many others, lie at the core of the most serious ecological dislocations we face today—apart, to be sure, from those that are produced by natural catastrophes.

Social ecology locates the roots of the ecological crisis firmly in relations of domination between people. The domination of nature is seen as a product of domination within society, but this domination only reaches crisis proportions under capitalism. In the words of Murray Bookchin, the main thinker of social ecological thought:

"The notion that man must dominate nature emerges directly from the domination of man by man... But it was not until organic community relations... dissolved into market relationships that the planet itself was reduced to a resource for exploitation. This centuries-long tendency finds its most exacerbating development in modern capitalism. Owing to its inherently competitive nature, bourgeois society not only pits humans against each other, it also pits the mass of humanity against the natural world. Just as men are converted into commodities, so every aspect of nature is converted into a commodity, a resource to be manufactured and merchandised wantonly. The plundering of the human spirit by the market place is paralleled by the plundering of the earth by capital." (Post Scarcity Anarchism)

## Planetary Phase of Civilization

Centuries earlier, civilizations were localized and while one went into ruination, another one in another part of the world thrived. Concept of the Planetary Phase of Civilization proposes that today's civilization is a global phenomenon due to increased global connectivity, new information technologies, environmental change in the biosphere, economic globalization, and shifts in culture and consciousness. While it is generally agreed that some form of planetary society is taking shape and binding the world's people and biosphere into a common fate, the future character of global society is uncertain and contested.

The concept of the planetary phase of civilization has become popular in the academic field of environmental science. In "Building a Global Culture of Peace," Steven C. Rockefeller states that "...we have entered a planetary phase in the development of civilization – what the historians call an era of global history." In his article entitled "Paths to Planetary Civilization," Ervin László describes this planetary civilization as one in which "The consensually created and globally coordinated ecosocial market system begins to function" and "The natural resources required for health and vitality become available to all the peoples and countries of the human community."

This kind of scenario analysis helps analysts think in an organized fashion about future alternatives, key decision points, and possible obstacles to global development. It then becomes possible to determine how to avoid the less-favorable directions and encourage changes to nurture a more beneficial one.

### Global Citizens Movement

The concept of a global citizen first emerged among the Greek Cynics in the 4th Century BC, who coined the term "cosmopolitan" – meaning citizen of the world. The Roman Stoics later elaborated on the concept. The contemporary concept of cosmopolitanism, which proposes that all individuals belong to a single moral community, has gained a new salience as scholars examine the ethical requirements of the planetary phase of civilization.

In most discussions, the global citizens movement is a socio-political process rather than a political organization or party structure. The term is often used synonymously with the anti-globalization movement, the movement of movements, or the global justice movement.

Global citizens movement has been used by activists to refer to a number of organized and overlapping citizens groups who seek to influence public policy often with the hope of establishing global solidarity on an issue. Such efforts include advocacy on ecological sustainability, corporate responsibility, social justice and similar progressive issues.

Today's objective and subjective conditions have led to emergence of a global civic identity and a latent pool of tens of millions of people ready to identify around new values of earth consciousness.

### Deep Ecology

The phrase deep ecology was coined by the Norwegian philosopher Arne Næss in 1973. Bill Devall and George Sessions, two prominent proponents, define Deep ecology, "as a process of ever-deeper questioning of ourselves, the assumptions of the dominant worldview in our culture, and the meaning and truth of our reality."

The core principle of deep ecology is that, like humanity, the living environment as a whole has the same right to live and flourish. Deep ecology is a recent branch of ecological philosophy (ecosophy) that considers humankind an integral part of its environment. Deep ecology has led to a new system of environmental ethics. Deep ecology describes itself as "deep" because it persists in asking deeper questions concerning "why" and "how" and thus is concerned with the fundamental philosophical questions about the impacts of human life as one part of the ecosphere, rather than with a narrow view of ecology as a branch of biological science.

"In technocratic-industrial societies there is overwhelming propaganda and advertising which encourages false needs and destructive desires designed to foster increased production and consumption of goods," say Devall and Sessions. "Most of this actually diverts us from facing reality in an objective way and from beginning the 'real work' of spiritual growth and maturity."

Deep ecologists would like to see much of the world returned to wilderness. They also speak of the "biocentric equality" of all living things. By this they mean that "all things in the biosphere have an equal right to live and blossom and to reach their own individual forms of unfolding and self-realization within the larger Self-realization."

Arne Næss rejected the idea that beings can be ranked according to their relative value. For example, judgments on whether an animal has an eternal soul, whether it uses reason or whether it has consciousness (or indeed higher consciousness) have all been used to justify the ranking of the human animal as superior to other animals. Næss states that "the right of all forms [of life] to live is an universal right which cannot be quantified. No single species of living being has more of this particular right to live and unfold than any other species."

Deep ecology offers a philosophical basis for environmental protection which may, in turn, guide human activity against perceived self-destruction. Deep ecology and environmentalism hold that the science of ecology shows that ecosystems can absorb only limited change by humans or other dissonant influences. Further, both hold that the actions of modern civilization threaten global ecological well-being.

Regardless of which model is most accurate, environmentalists contend that massive human economic activity has pushed the biosphere far from its "natural" state through reduction of biodiversity, climate change, and other influences. As a consequence, civilization is causing mass extinction. Deep ecologists hope to influence social and political change through their philosophy.

## Deep Ecology and Spirituality

The central spiritual tenet of deep ecology is that the human species is a part of the Earth and not separate from it. A process of self-realisation or "re-earthing" is used for an individual to intuitively gain an ecocentric perspective. The notion is based on the idea that the more we expand the self to identify with "others" (people, animals, ecosystems), the more we realise ourselves.

Other traditions which have influenced deep ecology include Taoism, Buddhism and Jainism primarily because they have a non-dualistic approach to subject and object. In relation to the Judeo-Christian tradition, Næss offers the following criticism: "The arrogance of stewardship [as found in the Bible] consists in the idea of superiority which underlies the thought that we exist to watch over nature like a highly respected middleman between the Creator and Creation." This theme had been expounded in Lynn Townsend White, Jr.'s 1967 article "The Historical Roots of Our Ecological Crisis", in which however he also offered as an alternative Christian view of man's relation to nature that of Saint Francis of Assisi, who he says spoke for the equality of all creatures, in place of the idea of man's domination over creation.

Proponents of deep ecology believe that the world does not exist as a resource to be freely exploited by humans. Present human interference with the nonhuman world is excessive, and the situation is rapidly worsening.

In practice, deep ecologists support decentralization, the creation of eco-regions and the breakdown of industrialism in its current form.

Parallels have been drawn between deep ecology and other movements, in particular the animal rights movement and 'Earth First' movement.

## Development Criticism

Development criticism, also known as anti-modernism, refers to criticisms of modern technology, industrialization, capitalism and economic globalization. Development critics see modernization as harmful for both humans and the environment.

Development criticism was born with the modern concept of development. One famous critic of modern life in the nineteenth century was the writer Henry Thoreau, who preferred living in the woods to living in the city.

The best-known development critic is Mohandas Gandhi, who heavily criticized modern technology and many other characteristics of western culture. Like many other development critics, he recommended local food production for local consumption rather than for trade. Similar thinkers often criticize contemporary globalization.

Happiness is a central theme of development-critical writings. Modern societies, despite their goal-oriented complexity and amount of labour time, do not help people to reach happiness, according to some development critics. In their view, happiness may be harder to reach in modern society than in primitive ones.

Often development critics criticize concepts used in modern societies, such as poverty and other welfare-related conceptualizations such as the human development index and gross national product. They say such concepts make the life of primitive or alternative societies look misleadingly dull to modern people. Modern societies apply subjective standards for welfare universally and (mis)judge other societies by them. Development critics often regard attempts to develop non-developed societies as a cause of misery and trouble, and thus recommend that development projects should be cancelled. Some even see the word "development" as negative and think that it represents conceptual imperialism.

## Ecotheology

Ecotheology is a form of constructive theology that focuses on the interrelationships of religion and nature in the light of environmental concerns. Ecotheology starts from the premise that a relationship exists between human religious/spiritual worldviews and the degradation of nature. It explores the interaction between ecological values, such as sustainability, and the human domination of nature. The movement has produced numerous religious-environmental projects around the world.

The relationship of theology to the modern ecological crisis became an intense issue of debate in Western academia in 1967, following the publication of the article, "The Historical Roots of Our Ecological Crisis, " by Lynn White, Professor of History at the University of California at Los Angeles. In this work, White puts forward a theory that the Christian model of human dominion over nature has led to environmental devastation.

Hindu ecotheology includes writers such as Vandana Shiva. Seyyid Hossein Nasr, a liberal Muslim theologian, was one of the earlier voices calling for a re-evaluation of the Western relationship to nature.

Christianity has often been viewed as the source of negative values towards the environment. Of course there are many voices within the Christian tradition whose vision embraces the well-being of the earth and all creatures. While St. Francis of Assisi is one of the more obvious influences on Christian ecotheology, there are many theologians and teachers whose work has profound implications for Christian thinkers. Many of these are less well-known in the West because their primary influence has been on the Orthodox Church rather than the Roman Catholic Church.

## Eco-communalism

Eco-communalism is an environmental philosophy based on ideals of simple living, local economies, and self-sufficiency. Eco-communalists envision a future in which the economic system of capitalism is replaced with a global web of economically interdependent and interconnected small local communities. Decentralized government, a focus on agriculture and 'green economics'.

One of the prominent documents on the subject describes eco-communalism as a "vision of a better life" which turns to "non-material dimensions of fulfillment – the quality of life, the quality of human solidarity and the quality of the earth."

Eco-communalism finds its roots in a diverse set of ideologies which include:

Pastoral reaction to industrialization-William Morris.

Nineteenth-century social utopians- Thompson, 1993.

The Small Is Beautiful philosophy-E.F. Schumacher, 1972. Traditionalism of Gandhi-Great Transition, 1993.

The term eco-communalism was first coined by the Global scenario group (GSG), which was convened in 1995 by Paul Raskin.

It urges our society to advance past reckless industrialism towards a more localized, environmentally palatable system.

In 1983, E.F. Schumacher published Small Is Beautiful, a collection of essays in which he expressed the unsustainability of the modern world's consumption behavior and the need for a new outlook to prevent otherwise inevitable environmental collapse: "Ever bigger machines, entailing ever bigger concentrations of economic power and exerting ever greater violence against the environment, do not represent progress: they are a denial of wisdom. Wisdom demands a new orientation of science and technology towards the organic, the gentle, the non-violent, the elegant and beautiful."

Rather than a world of capitalist states and their often exploited workers driven by their own greed, eco-communalism envisions a world in which government is decentralized, settlements are integrated with larger cities, local farming is the primary source of produce, and ecological thinking and interconnectedness are the new human values.

### Amish

The Amish are an Anabaptist Christian denomination, formed in 1693 by Swiss Mennonites led by Jacob Amman. They live in the United States and Canada and are divided into several major groups. The Old Order Amish use horses for farming and transportation, dress in a traditional manner, and forbid electricity or telephones in the home. Church members do not join the military, apply for Social Security benefits, take out insurance or accept any form of financial assistance from the government.

At home, most Amish speak a dialect known as Pennsylvania Dutch, Pennsylvania German, or Deitsch. Children learn English in school. The Amish are divided into separate fellowships consisting of geographical districts or congregations. Each district is fully independent and has its own Ordnung, or set of unwritten rules. Old Order churches may shun or expel members who violate these rules. Amish mode of transport is mostly horse wagons and they avoid cars.

## Human Ecology

Human ecology is an academic discipline that deals with the relationship between humans and their natural, social and created environments. Human ecology investigates how humans and human societies interact with nature and with their environment.

Human Ecology, as an interdisciplinary applied field, uses a holistic approach to help people solve problems and enhance human potential within their near environments - their clothing, family, home, and community. Human Ecologists promote the well-being of individuals, families, and communities through education, prevention, and empowerment.

In the USA, human ecology was established as a sociological field in the 1920's, although geographers were using the term much earlier. Amos H. Hawley published 'Human Ecology - A Theory of Community Structure' in 1950.

Humans are no longer seen as an exceptional species that uses culture to adapt to new environments and environmental change, influenced more by social than by biological variables, but rather as one species out of many that interacts with a bounded natural environment.

Influences can be seen between human ecology and the field of political ecology. Human ecology explores not only the influence of humans on their environment but also the influence of the environment on human behaviour, and their adaptive strategies as they come to understand those influences better.

## Voluntary Human Extinction Movement

The Voluntary Human Extinction Movement, or VHEMT (pronounced "vehement"), is a movement that calls for the voluntary extinction of the human race with a motto, "May we live long and die out".

The basic concept behind VHEMT is the belief that the Earth would be better off without humans, and as such, humans should refuse to breed. However, this does not mean that they intend to force people to not breed, to kill anyone, or to commit suicide.

Les U. Knight, credited with giving the name "Voluntary Human Extinction Movement", is the owner of VHEMT.org, and is cited as the founder, de facto leader or "prime avatar", in different publications. VHEMT Volunteers refuse to have no children (or no more, if they already have children).

In 2001, Knight appeared on Hannity & Colmes to present VHEMT's ideology. On the program, he stated that "as long as there's one breeding pair of homo sapiens, there's too great a threat to the biosphere." He also expressed no hope for voluntary human extinction, but stated that "it is the right thing to do."

VHEMT spreads its message through the Internet, thus reaching mainly wealthier nations. As a general rule, these countries already have fertility rates below the replacement rate and are thus already trending toward "human extinction," or at least a reduced population. However, according to VHEMT, wealthier nations have the largest impact on world resources.

## Homesteading

Broadly defined, homesteading is a lifestyle of simple, agrarian self-sufficiency.

Currently the term homesteading applies to anyone who is a part of the back-to-the-land movement and who chooses to live a sustainable, self-sufficient lifestyle. While land is no longer freely available in most areas of the world, homesteading remains as a way of life. A new movement, called "urban homesteading," can be viewed as a simple living lifestyle, incorporating small-scale agriculture, sustainable and permaculture gardening, and home food production and storage into suburban or city living.

## Ecotivity

Ecotivity is an abbreviation of the term, ecological activity. It denotes any activity that promotes the conservation or sustainability of ecosystems and bio-diversity.

In more general terms, an ecotivity can be thought of as any ecologically friendly activity. Ecotivities therefore make a positive contribution to averting the environmental crisis of the planet or means of effecting positive change.

Ecotivities include all associated values, actions, and choices either as a green consumer, or as an individual living a sustainable lifestyle, for example, by recycling or energy efficiency. Ecotivity can also be of political, direct action, hand-on environmental conservation, and creative confrontational acts, that are a means to protect the environment.

## Self-sufficiency

### Consume What You Produce, Produce What You Need

Self-sufficiency is the state of not requiring any aid, support, or interaction, for survival; it is therefore a type of personal or collective autonomy. On a national scale, a totally self-sufficient economy that does not trade with the outside world is called an autarky.

The term self-sufficiency is usually applied to varieties of sustainable living in which nothing is consumed outside of what is produced by the self-sufficient individuals. Practices that enable or aid self-sufficiency include autonomous building, permaculture, sustainable agriculture, and renewable energy.

### Post-modern self-sufficiency

The term 'post-modern self-sufficiency' or 'escape capitalism' refers to a mode of life that seeks to exist outside industrialized non-agrarian 'Western' norms.

## Earth Day

Earth Day was first conceived by John McConnell as a global holiday to celebrate the wonder of life on our planet and to inspire awareness of and appreciation for the Earth's environment. The United Nations celebrates Earth Day each year on the March equinox. Also an observance originated by Gaylord Nelson as an environmental teach-in, is also being called Earth Day since 1970 and celebrated in many countries on April 22.

March Equinox marks the first moment of Spring, when day and night are equal around the world and hearts and minds can join together with thoughts of harmony and Earth's rejuvenation. Just as a single prayer can be significant, how much more so when hundreds, thousands, millions of people throughout the world join in peaceful thoughts and prayers to nurture neighbor and nature.

The first Proclamation of Earth Day was by San Francisco, the City of Saint Francis, patron saint of ecology. Designating the First Day of Spring, March 21, 1970 to be Earth Day, this day of nature's equipoise was later sanctioned in a Proclamation signed by Secretary General U Thant at the United Nations where it is observed each year. Earth Day was firmly established for all time on a sound basis as an annual event to deepen reverence and care for life on our planet.

On April 22 in year 1970, 20 million Americans took to the streets, parks, and auditoriums to demonstrate for a healthy, sustainable environment. Denis Hayes, the national coordinator, and his youthful staff organized massive coast-to-coast rallies. Thousands of colleges and universities organized protests against the deterioration of the environment. Groups that had been fighting against oil spills, polluting factories and power plants, raw sewage, toxic dumps, pesticides, freeways, the loss of wilderness, and the extinction of wildlife suddenly realized they shared common values.

Mobilizing 200 million people in 141 countries and lifting the status of environmental issues onto the world stage. Earth Day on April 22 in 1990 gave a huge boost to recycling efforts worldwide and helped pave the way for the 1992 United Nations Earth Summit in Rio de Janeiro.

Margaret Mead added her support for the equinox Earth Day, and in 1978 declared:

"Earth Day is the first holy day which transcends all national borders, yet preserves all geographical integrities, spans mountains and oceans and time belts, and yet brings people all over the world into one resonating accord, is devoted to the preservation of the harmony in nature and yet draws upon the triumphs of technology, the measurement of time, and instantaneous communication through space."

# Solution -20

## Simple Living High Thinking
## Communities

*If it is true that the Age of Growth is over, and the Age of Entropy has begun, and if we are to retain any hope of a reasonable quality of life without destroying other people's, then our infrastructure, our settlements, our industries and our lives require total reconstruction.*

*– George Monbiot*

Our dependence on oil grew over just 150 years. That's a mere speck in time in the history of our world. This chapter is about downsizing when we can, which is better than doing it when we must.

The whole world is trying to imitate western lifestyle of extravagance and in the long run that has no future. Based on cheap oil we have developed an expansion economy in a now too full world. Industrialized life style is the greatest misallocation of resources in the history of the world. The peak of oil production should also be the peak of globalization.

Looks like localized self-sufficient communities are the way to go.

Once again take the case of North America where the supply of natural gas is rapidly disappearing. In a reflection of this desperate (and demented) condition, Canada is now starting to divert some of its remaining natural gas to the manufacture of synthetic oil from tar sands, so as to ease the pressure on supplies of conventional petroleum. Given the prohibitive cost of building gas pipelines from Asia and Africa, the only practical way to get more gas supplies to North America would be to spend several hundred billion dollars (or more) on facilities for converting foreign sources of gas into liquified natural gas (LNG), shipping the LNG in giant doubled-hulled vessels across the Atlantic and Pacific, and then converting it back into a gas in "regasification" plants in American harbors. Although favored by the Bush administration, plans to construct such plants have provoked opposition in many coastal communities because of the risk of accidental explosion as well as the potential for inviting terrorist attacks.

Why make things so complicate? Why not encourage people to adapt to simpler and more peaceful life? Otherwise where will all these unending complex diplomacies and stopgap solutions will lead us to.

We seem to have a tendency to deal with problems on an ad hoc basis - i.e., to deal with 'problems of the moment'. This does not foster an attitude of seeing a problem embedded in the context of another problem.

Globalisation makes it impossible for modern societies to collapse in isolation. Any society in turmoil today, no matter how remote, can cause problems for prosperous societies on other continents, and is also subject to their influence (whether helpful or destabilising).

George Monbiot rightly puts it, "I know this is also the way forward for the world. The urban-industrial complex as we know it will one day disappear. We'll ultimately run out of oil. When this happens, the face of society will change. Then people will feel more dependent on powers beyond them, and to many that will mean developing a sense of God. But when the oil runs out, we don't want to be left high and dry."

## Self-Sufficient, Localized, God Centred Communities

In this section, we will study a few self-sufficient communities.

Way back in 1974, Srila Prabhupada wrote:

"Our farm projects are an extremely important part of our movement. We must become self-sufficient by growing our own grains and producing our own milk. *There will be no question of poverty. They should be developed as an ideal society* dependent on natural products, not industry." *(Srila Prabhupada, Letter dated December 18, 1974)*

## New Vrindaban

New Vrindaban is an ISKCON (Hare Krishna) rural community located in Moundsville, West Virginia. It functions as a spiritual pilgrimage center which attracts people from all over the world, and also as a community striving to enact a model of self-sufficiency based on spiritual ideals and practices. New Vrindaban is named after the Indian town of Vrindavan.

The community was founded in 1968 by His Divine Grace A.C.Bhaktivedanta Swami Prabhupada to create an atmosphere of

"Simple Living and High Thinking". New Vrindaban is meant to present to the rest of the world a return to Vedic village living, depending on the land, the cow, and ultimately upon the Supreme Personality of Godhead, Krishna. This community is meant to be based on God consciousness, simplicity, enlightened agriculture, cow protection, and ox power.

New Vrindaban is a strictly vegetarian community, in order to facilitate spiritual growth. It is believed in the philosophy of Krsna Consciousness that reactions due to meat consumption create negative karma. The intention was also to set a proper example of sustainability to other similar intentional communities and to the general public.

After few years of ups and downs and controversies, New Vrindaban today is one of ISKCON's most well-known and well-visited (over 40,000 visitors a year) temples in North America.

Devotees stay in a three-story temple/ashram, where they practice devotional activities to the presiding Deities, Sri-Sri Radha Vrindaban Chandra, who were installed at New Vrindaban on Janmastami, August 13, 1971. These activities of devotional service include kirtan (congregational chanting of the Holy Names of the Lord, and the cooking and offering of prasadam, vegetarian foodstuffs offered to the Deities. It is believed this process allows for a subtle spiritual essence to enter the foodstuffs to aid the spiritual advancement of anyone who partakes of it. Other activities include pujari, or priestly duties done for the Deities (like dressing, bathing, offering of different articles like incense), fundraising, preaching to visitors, temple and grounds maintenance, and spiritual educational programs.

The large acreage on the New Vrindaban property also supports a number of agrarian projects. Over 80 cows are kept in a protection program that allows them to live out their natural life without being slaughtered. Two agricultural projects, the S.A.N.T.E.E. (The

*"Every time I see an adult on a bicycle, I no longer despair for the human race. It's a start, but I'd feel even more confident about our chances of survival if I saw George Bush and Dick Cheney sharing a car to work."*
*- HG Wells*

Sustainable Agricultural Network for Training and Environmental Education) Teaching Garden (1 acre) and the Garden of Seven Gates (6.5 acres) provides the community with locally-grown organic produce, herbs, and other natural products.

Devotees at New Vrindaban also travel to local universities to do vegan/vegetarian cooking classes and seminars on Krishna consciousness and other spiritual topics and lifestyle choices. A few devotees based at New Vrindaban also travel across America distributing the books of Prabhupada and other Krishna conscious literature.

We can mention couple of more communities in Europe which are based on the principle of voluntary simplicity, of course without a spiritual theme at their core.

## Vauban Community, Germany

### Car Free Living

It's pickup time at the Vauban kindergarten here at the edge of the Black Forest, but there's not a single minivan waiting for the kids. Instead, a convoy of helmet-donning moms - bicycle trailers in tow - pedal up to the entrance.

Welcome to Germany's best-known environmentally friendly neighborhood and a successful experiment in green urban living. The Vauban development - 2,000 new homes on a former military base 10 minutes by bike from the heart of Freiburg - has put into practice many ideas that were once dismissed as eco-fantasy but which are now moving to the center of public policy.

With gas prices well above $6 per gallon across much of the continent, Vauban is striking a chord in Western Europe as communities encourage people to be less car-dependent.

"Vauban is clearly an offer for families with kids to live without cars," says Jan Scheurer, an Australian researcher who has studied the Vauban model extensively. "It was meant to counter urban sprawl - an offer for families not to move out to the suburbs and give them the same, if better quality of life. And it is very successful."

There are numerous incentives for Vauban's 4,700 residents to live car-free: Carpoolers get free yearly tramway passes, while parking spots - available only in a garage at the neighborhood's edge - go for 17,500 (US$23,000). Forty percent of residents have bought spaces, many just for the benefit of their visiting guests.

As a result, the car-ownership rate in Vauban is only 150 per 1,000 inhabitants, compared with 430 per 1,000 inhabitants in Freiburg proper.

In contrast, the US average is 640 household vehicles per 1,000 residents. But some cities - such as Davis, Calif., where 17 percent of residents commute by bike - have pioneered a car-free lifestyle that is similar to Vauban's model.

Vauban, which is located in the southwestern part of the country, owes its existence, at least in part, to Freiburg - a university town, like Davis - that has a reputation as Germany's ecological capital.

In the 1970s, the city became the cradle of Germany's powerful antinuclear movement after local activists killed plans for a nuclear power station nearby. The battle brought energy-policy issues closer to the people and increased involvement in local politics. With a quarter of its people voting for the Green Party, Freiburg became a political counterweight in the conservative state of Baden-Württemberg.

At about the same time, Freiburg, a city of 216,000 people, revolutionized travel behavior. It made its medieval center more pedestrian-friendly, laid down a lattice of bike paths, and introduced a flat rate for tramways and buses.

Environmental research also became a backbone of the region's economy, which boasts Germany's largest solar-research center and an international center for renewable energy. Services such as installing solar panels and purifying wastewater account for 3 percent of jobs in the region, according to city figures.

*"I believe the long-term solution requires nothing less than the gradual abandonment of the lethal techniques, the uncongenial lifeways, and the dangerous mentality of industrial civilization itself. This would imply the end of the giant factory, the huge office, perhaps of the urban complex."*
*-Economist Robert Heilbroner*

Little wonder then, that when the French Army closed the 94-acre base that Vauban now occupies in 1991, a group of forward-thinking citizens took the initiative to create a new form of city living for young families.

Across Europe, similar projects are popping up. Copenhagen, for instance, maintains a fleet of bikes for public use that is financed through advertising on bicycle frames.

But what makes Vauban unique, say experts, is that "it's as much a grass-roots initiative as it is pursued by the city council," says Mr. Scheurer. "It brings together the community, the government, and the private sector at every state of the game."

As more cities follow Vauban's example, some see its approach taking off. "Before you had pilot projects. Now it's like a movement," says Mr. Heck. "The idea of saving energy for our landscape is getting into the basic planning procedure of German cities."

*Copyright © 2006 The Christian Science Monitor*

## Tinkers Bubble Community

### Somerset, England

Tinkers' Bubble is 40 acres of woodland, orchards and pasture in south Somerset, England. It was bought by a group of environmentalists in 1994, and a dozen people moved in, applied for shares and built themselves temporary houses.

They imposed a strict set of rules on themselves, which included a ban on the use of internal combustion engines on the land. They made a partial exception for transport: the 12 residents share two cars.

Otherwise, the only fossil fuel they consume is the paraffin they put in their lamps. They set up a small windmill and some solar panels, built compost toilets, and bought a wood-powered steam engine for milling timber, some very small cows and a very large horse.

The first winter was spent wading around in two feet of mud. Some of the locals, mistaking the settlers for new age travellers, went berserk. There was plenty of internal strife as well.

The work is tough. They fell trees with handsaws, heat their homes with wood, cut the hay with scythes and milk the cows, weed the fields and harvest the crops by hand.

But they have come through. They have made friends with the locals, who are coming to see the project as an asset: the land is biodiverse, still has standing orchards, and is open to the public.

Their stall has won first prize in the local farmers' market. They have learned, often painfully, to live together. Because it doesn't depend on heavy machinery, this farm, unlike most, isn't in hock to the bank.
- George Monbiot

# Appendix

Hare Krishna Farm Communities
Around The World

## NORTH AMERICA
## CANADA
Ashcrot, B.C.-Saranagati Dhama, Box 99, Ashcrott, B.C. V0K 1A0
## USA
Alachua, Florida (New Ramana-reti)-Box 819, Alachua. 32615/ Tel. (904) 462-2017

Carriere, Mississippi (New Talavan)-31492 Anner Road, 39426/ Tel. (601 ) 798-6623

Gurabo, Puerto Rico (New Govardhana Hill)-(contact ISKCON Gurabo)

Hillsborough, North Carolina (New Goloka)-Rt. 6, Box 701, 27278/ Tel. (919) 732-6492

Mulberry, Tennessee (Murari-sevaka)-Rt. No. 1, Box 146-A. 37359/ Tel (615) 759-6888

Port Royal, Pennsylvania (Gita Nagari)-R.D. No. 1, Box 839, 17082/Tel. (717) 527-4101

## EUROPE
## UNITED KINGDOM AND IRELAND
Lisnaskea, North Ireland-Lake Island of Inis Rath, Lisnaskea Co. Fermanagh/ Tel. +44 (03657) 21512

London, England-(contact Bhaktivedanta Manor)
## GERMANY
Jandelsbrunn-Nava-Jiyada-Nrsimha-Ksetra, Zielberg 20, W-8391 Jandelsbrunn/ Tel +49 85831332
## ITALY
Florence (Villa Vrindavan)-Via Communale degli Scopeti 108, S. Andrea in Percussina, San Casciano, Val di Pesa (Fl) 5002/ Tel. +39 (055) 820-054

## SWEDEN
Jarna-Almviks Gard, 15300 Jarna / Tel. +46 (08) 551-52050; 551-52073
## OTHER COUNTRIES
Czech Republic-Krsnuv Dvur c. 1, 257 28 Chotysany

Denmark-Gl. Kirikevej 3, 6650 Broerup/ Tel. +45 (075) 392921

France (La Nouvelle Mayapura)-Domaine d'Oublaisse, 36360, Lucay le Male/ Tel. +33 (054) 402481

Poland (New Santipura)-Czarnow 21, k. Kamiennej gory, woj. Jelenia gora / Tel. +48 8745-1892

Spain (New Vraja Mandala)-(Santa Clara) Brihuega, Guadalajara / Tel. +34 (911 ) 280018

Switzerland-Gokula Project, Vacherie Dessous, 2913 Roche d'Or/ Tel. +41 (066) 766160

## AUSTRALASIA
## AUSTRALIA
Bambra (New Nandagram)-Oak Hill, Dean s Marsh Road, Bambra, VIC 3241/ Tel +61 (052) 88-7383

Millfield, N.S.W.-New Gokula Farm, Lewis Lane (off Mt.View Rd. Midfield near Cessnock),

N.S.W. (mail: P.O. Box 399, Cessnock 2325, N.S.W., Australia)/ Tel. +61 (049) 98-1800

Murwillumbah (New Govardhana)-Tyalgum Rd., Eungella, via Murwillumbah N. S. W. 2484 (mail: P.O. Box 687)/ Tel. +61 (066) 72-1903
## NEW ZEALAND AND FIJI
Auckland, New Zealand (New Varshan)-Hwy. 18, Riverhead, next to Huapai Golf Course (mail: R.D. 2, Kumeu, Auckland)/ Tel. +64 (09) 4128075
## AFRICA
Mauritius (ISKCON Vedic Farm)-Hare Krishna Rd., Vrindaban, Bon Acceuil/ Tel. 418-3955

## ASIA
## INDIA
Ahmedabad District, Gujarat-Hare Krishna Farm, Katwada (contact ISKCON Ahmedabad)

Assam-Karnamadhu, Dist. Karimganj

Chamorshi, Maharashtra-78 Krishnanagar Dham, District Gadhachiroli, 442 603

Hyderabad, A. P.-P. O. Dabilpur Village, Medchal Tq., R.R. District, 501 401/ Tel. 552924

Mayapur, W. Bengal-(contact ISKCON Mayapur)

OTHER COUNTRIES

Indonesia-Govinda Kunja (contact ISKCON Jakarta)

Malaysia-Jalan Sungai Manik, 36000 Teluk Intan, Perak

Phillippines (Hare Krishna Paradise)-231 Pagsabungan Rd., Basak, Mandaue City/ Tel. +63 (032) 83254


LATIN AMERICA

BRAZIL

Pindamonhangaba, SP-Nova Gokula, Bairro de Ribeirao Grande (mail: C.P. 108, CEP 12400000)/ Tel. +55 (0122) 42-5002

Caruaru, PE-Nova Vrajadhama, Distrito de Murici (mail: C.P. 283, CEP 55000-000)

Parati, RJ-Goura Vrindavana, Sertao Indaituba (mail: 62 Parati, CEP 23970-000

MEXICO

Guadalajara-Contact ISKCON Guadalajara

PERU

Hare Krishna-Correo De Bella Visla-DPTO De San Martin

OTHER COUNTRIES

Argentina (Bhaktilata Puri)-Casilla de Correo No 77, 1727 Marcos Paz, Pcia. Bs. As., Republica Argentina

Bolivia-Contact ISKCON Cochabamba

Colombia (Nueva Mathura)-Cruzero del Guali, Municipio de Caloto, Valle del Cauca/ Tel. 612688 en Cali

Costa Rica-Granja Nueva Goloka Vrindavana, Carretera a Paraiso, de la entrada del Jardin Lancaster (por Calle Concava), 200 metros as sur (mano derecha) Cartago (mail: Apdo. 166, 1002)/ Tel. +506 51-6752

Ecuador(Nueva Mayapur)-Ayampe (near Guayaquil)

El Salvador-Carretera a Santa Ana, Km. 34, Canton Los Indios, Zapotitan, Dpto. de La Libertad

Guvana-Seawell Village, Corentyne, East Berbice

# The Global Ecovillage Directory

(For further information on the entries here, please visit: http://directory.ic.org/records/ecovillages.php)

New Earth Mountain Village (British Columbia, Canada) Forming

1st Sustainable Eco-Camp Settlement Project & International Research Center (Eastern Merume Mountains / Mazaruni and Kamarang Rivers, Cuyuni/Mazaruni / Potaro-Siparuni, Guyana) Forming

5th dimensional eco-village (Zuni Mountains, New Mexico, United States)

8th Life EcoVillage (La Palma, Canary Islands, Spain) Forming

Abeo (Hobart, Tasmania, Australia) Re-Forming

ABRA144 Ecovillage - Amazonian Bioregional Village (Presidente Figuereido, Manaus, Brazil) Forming

Abundance EcoVillage (Fairfield, Iowa, United States)

Aguas segrada (barucito, perez zeledon, Costa Rica) Forming

The Alchemical Nursery Project (Syracuse, New York, United States) Forming

Aldea 506 Ecovillage (San Carlos, Costa Rica) Forming

ALDEAFELIZ (San Francisco, Cundinamarca, Colombia)

Aldeia Arawikay (Antonio Carlos, Santa Catarina, Brazil) Forming

Aldinga Arts EcoVillage (Aldinga Beach, South Australia, Australia)

Aleskam (684020 Razdolny Setl, Kamchatka, Russian Federation)

Amity Highlands Ecovillage Cohousing (Woodbridge, Connecticut, United States) Forming

Anam Cara Community (Wales, United Kingdom) Forming

Andelssamfundet i Hjortshøj (DK-8530 Hjortshoj, Denmark)

Ant Hill Collective! (San Diego, California, United States)

Apasana (Austin, Texas, United States) Forming

Apusenii Verzi Ecovillage (Nadastia, Alba, Romania)

Aquarius Nature Retreat (Vail, Arizona, United States) Re-Forming

Arca Verde (Sao Francisco de Paula, Rio Grande do Sul, Brazil)

Arcosanti (Cordes Lakes, Arizona, United States)

Ardhanariswaras (haiku, Hawaii, United States) Forming

ASL in Veg/an Ecovillage Communities (Arizona, United States) Forming

Aspenwood (Santa Fe, New Mexico, United States)

Associação Ecológica Portal do Sol (São Francisco de Paula, Rio Grande do Sul, Brazil) Forming

Auro-Ecovillage Shawnigan Lake - British Columbia (Shawnigan Lake, British Columbia, Canada) Forming

Austin Area Ecovillages (Austin (area), Texas, United States) Forming

Avalon Community (Florida, United States) Forming

Awaawaroa Bay Eco-Village (Waiheke Island, Auckland, New Zealand)

Awakened Life Project (Benfeita, Portugal)

A-WAY-KIN ECOvillage (Gimli, Manitoba, Canada)  Forming

A-Woodland-Institute for Strategic Ecology (Lexington, South Carolina, United States) (Pelion, South Carolina, United States)  Forming

Back-to-Nature EcoVillage (Trang, Thailand)  Forming

Backyard Neighborhoods (San Francisco, California, United States) (and any other city)  Forming

Ballintaggart on the Atherton Tableland (Malanda, Queensland, Australia)

Base Camps Conservation Network (ARATULA, Queensland, Australia)  Forming

Bay View Ecovillage (Milwaukee, Wisconsin, United States)

Belfast Cohousing & Ecovillage (Belfast, Maine, United States)  Forming

Bellbunya Community Association (Belli Park, Queensland, Australia)

Berea College Ecovillage (Berea, Kentucky, United States)

Berkeley California Vegan Community (Berkeley, California, United States)  Forming

Better In Belize (South-east of Benque Viejo Del Carmen, Cayo District, Belize)  Forming

Betterfields Community Development (United States) (United States)  Forming

Between Two Cities (Portland, Corvallis, Oregon, United States)  Forming

Bhrugu Aranya Ecovillage - Poland (Wysoka-Jordanow, Malpolska, Poland)

Big Island Cohousing and Ecovillage (Hawaii, United States)

Biospharms Christian retreats (Cayo, Belize)  Forming

Birds and Bees Permaculture Village (Viroqua, Wisconsin, United States)  Forming

Bloomington Cooperative Plots Eco-Village (Bloomington, Indiana, United States)  Forming

The Blueberry Patch (Gulfport, Florida, United States)  Forming

Brave New Mountain (Hoodsport, Washington, United States)  Forming

Braziers Park (Wallingford, England, United Kingdom)

Breitenbush Hot Springs (Detroit, Oregon, United States)

Broken Earth Village (Lefke, New Mexico, Cyprus)

BTR Ecovillage (Dale (an hour outside of Austin, 15 minutes to Lockhart and Bastrop), Texas, United States)

Buckeye Permaculture co-op and eco-village (Buckeye, Arizona, United States)  Forming

Camelot Cohousing (Berlin, Massachusetts, United States)

CampOma (Eagle Creek 25 miles outside of Portland, Oregon, United States)  Forming

Carolina Common Well (Efland, North Carolina, United States)  Forming

Casa da Ribeira (Coimbra, Portugal)

Catholic Ecovillage (Kentucky, United States)  Forming

Centre de Ressource au Coeur de l'Etre (En Estrie, Québec) (Shefford, Quebec, Canada)  Forming

Cerro Gordo Ecovillage (Cottage Grove, Oregon, United States)

Chico Ecovillage (Chico, California, United States)  Forming

Christian Ecovillage-Mid-Atlantic Region (Bethlehem, Pennsylvania, United States)

Christie Walk (Adelaide, South Australia, Australia) (Christie Walk / Urban Ecology Australia 105 Sturt St Adelaide, SouthAustralia, Australia)

Chuckelberry Galactic Farms and Commodities (FairGrove, Missouri, United States)  Forming

Church of the Earth (Haiku, Hawaii, United States) (Hilo, Hawaii, United States) Forming

Cinderland Ecovillage (Kapoho, Hawaii, United States)

Circle of One (Phoenix, Arizona, United States) Forming

CIRCLE P ECOFARM/VILLAGE (brunswick, Georgia, United States) Forming

Cite Ecologique of New Hampshire (Colebrook, New Hampshire, United States) (Ham-Nord, Quebec, Canada)

Clear Water Sanctuary (Elma, Washington, Olympic Penninsula, United States) Forming

Cocomo Village (Hunga Island, Vava'u Group of Islands, Tonga, Neiafu, Vava'u, Tonga) Forming

CoHo Ecovillage (Corvallis, Oregon, United States)

Co-Housing Connection of East Hawaii (CCEHa) (Keaau, Hawaii, United States) (Hilo, Hawaii, United States) Forming

Columbia Ecovillage (Portland, Oregon, United States)

The Commons Ecovillage (Victoria, British Columbia, Canada) Forming

CommonUnity (Mullumbimby, New South Wales, Australia) Forming

Community Alive at Earthome (Pulteney, New York, United States) Forming

Community in Southern Saratoga County, NY (New York, United States) Forming

Community of Healers (Ash Fork, Arizona, United States) Forming

Comunidad Permacultural Na Lu' Um (San Cristobal de las Casas, Chiapas, Mexico) Forming

Comunidad Planetaria Janajpacha (Bolivia)

Concord Ecovillage (Kennett Square, Pennsylvania, United States) (Newark, Delaware, United States) (West Chester, Pennsylvania, United States) Forming

Costa Rica Ecovillage (Costa Rica) Forming

Create An Eden (New YorkCatskills Region, United States) Forming

Crystal Triple Moon Tribe (St.Johns, Arizona, United States) Forming

Crystal Waters Permaculture Village (Conondale, Queensland, Australia)

Dallas Cohousing / Ecovillage (Dallas, Texas, United States) Forming

Damanhur, Federation of Communities (Baldissero Canavese, Turin, Italy)

Dancing Rabbit Ecovillage (Rutledge, Missouri, United States)

Dapala Farm (Elk is located in Eastern Washington State Region, Washington, United States) Forming

Dedetepe eco-farm (Canakkale, Turkey) Forming

Deep River CoHousing at Living Well Community (Franklinville, North Carolina, United States) Forming

DFW Urban Ecovillage (East of Dallas suburbs, Texas, United States) Forming

Dolphin Community (D-79737 Herrischried - Niedergebisbach, Germany) (Hausen im Wiesental, Germany)

Dome Village Katrina (Mount Shasta, California, United States) Forming

Dragon Belly Farm (Pt. Ludlow, Washington, United States) Forming

Dreamtime Village (La Farge, Wisconsin, United States)

Earth Energies Eco Village (Marigot, St. Andrew, Dominica) Forming

Earth Kind Developments Corp. (Tofino, British Columbia, Canada) Forming

EarthArt Village (Moffat, Colorado, United States) Forming

Earthaven Ecovillage (Black Mountain, North Carolina, United States)

EarthChild Collective (Pollock Pines, California, United States) Forming
Earthsong Eco-Neighbourhood (Ranui, West Auckland, New Zealand)
Earthworks Eco Village (Madoc, Ontario, Canada) (Colborne, Ontario, Canada) Forming
EARTHYACHT (Mazatlan Sinaloa Northern Mexico Area, Mazatlan, Mexico) Forming
East Texas Intentional Earthship Community (Bullard, Texas, TX, United States) Forming
Eco Island (Smithland, Kentucky, United States) Forming
Eco Yoga Park (General Rodriguez, Buenos Aires, Argentina) (General Rodriguez, Buenos Aires, Argentina)
Ecoaldea Huehuecoyotl (Tepoztlán, Morelos, Mexico)
Ecoculture Village (Arcata, California, United States) Forming
EcoLetu (Beijing, China) Forming
Ecolivom City (Burlington, Massachusetts, United States) Forming
Economic Eco-Village (Glastonbury, Connecticut, United States) (Bellefonte Chestnut, Pennsylvania, United States) Forming
EcoReality (Salt Spring Island, British Columbia, Canada) Re-Forming
Ecotopia Romania (Stanciova, Timis, Romania) Forming
Ecotopia/Ithaca (Vathi/Ithaca, Greece) Forming
Ecovila Clareando (Piracaia, SP, Brazil) Forming
Ecovila da Montanha (Brazil) Forming
Ecovilla Asociación GAIA (NAVARRO, Buenos Aires, Argentina)
Ecovillage at Burdautien (Eire, Ireland)
The Ecovillage at Currumbin (Currumbin Valley, Queensland, Australia)
EcoVillage at Ithaca (Ithaca, New York, United States)
EcoVillage at Ithaca, TREE, the third neighborhood (Ithaca, New York, United States) Forming
Eco-Village de la Paix-Dieu (Jehay, Belgium)
EcoVillage Detroit (Detroit, Michigan, United States) Forming
Ecovillage in planning (Georgia, United States) Forming
Ecovillage Network UK (Bristol BS99 3JP, England, United Kingdom)
Eco-Village of La Hermita (San Miguel de Allende, GTO, 37700, Mexico)
EcoVillage of Loudoun County (Lovettsville, Virginia, United States) Forming
Écovillage 2012 (Montreal, Quebec, Canada) Forming
ECOVILLAGE VIVER SIMPLES (Itamonte, Minas Gerais, Brazil) Re-Forming
The Eden Project Forming
Edinstvo ecovillage (Edinstvo gardens, Kagalnitskiy rayon, Rostov Region, Russian Federation)
Ekobius (Kamarang, Cuyuni-Mazaruni, Guyana)
Eliopoli (Port Maitland, Nova Scotia, Canada) Re-Forming
ENARGEIA (Pilio, Greece)
Enota Mountain Village (Hiawassee, Georgia, United States)
EPIC (United States) Forming
Esperanza de Sol de Finca Amanecer (Londres de Quepos, Costa Rica) Forming
Etherion Ringing Cedars Intentional Community (Guyra, New South Wales, Australia) (Upwey, Victoria, Australia) Forming

Factor 'e' Farm - Open Source Ecology (Maysville, Missouri, United States) Forming
Family Haven Project (Nelson, Nelson District, New Zealand) Forming
The Farm (Summertown, Tennessee, United States)
farmvilleinreallife (Tampico, Mexico) Forming
Fedorovtsy (Staraya Tishanka, province of Voronezh, Russian Federation)
Finca Fruicion (Perez Zeledon, San Jose, Costa Rica) Forming
Findhorn Foundation and Community (Findhorn, Scotland, United Kingdom)
The Fire Circle (Russell or Heath, Massachusetts, United States) Forming
Flying Fish Organic Village (near Vuaki, Yasawa Islands Group, western Fiji, Blue Lagoon, Yasawa Islands, Fiji)
Footprint Acres (Lower Lake, California, United States) Forming
Forest Haven (Moyock, North Carolina, United States) Forming
Forest Moon Ecovillage (Greenville County, South Carolina, United States) Forming
Forgotten (Rocky Mountain National Forest, Colorado, United States) Forming
FOUNDATION (London, United Kingdom) (Tamil Nadu, India) Forming
Freesteading Project (Anononia, Córdoba, Argentina) Forming
Frieden Haven (Steinbach, Manitoba, Canada) Forming
Friendly Haven Rise Farm Ecovillage (Battle Ground, Washington, WA, United States) Forming
Front Range Eco Town (boulder, Colorado, United States) (denver, Colorado, United States) Forming
Gaia Grove Ecovillage (near Gainesville, Florida, United States)
Gaia Sangha (LOS ANGELES, California, United States) Forming
Gaia Shifts (Nelson BC, British Columbia, Canada)
Gaia University (Online)
Gaiam One (Chimirol De Rivas, San Jose, Costa Rica) Re-Forming
The Gaian Fellowship (Eureka Springs, Arkansas, United States) Forming
Gaia's Green Village (Lonedell, Missouri, United States) Forming
Gaia's Sanctuary (Boulder, Colorado, United States) Forming
Gaiasana (Turrialba, Cartago, Costa Rica) (Sandiago de Puriscal, SJ, Costa Rica) Forming
Global Community Communications Alliance (Tubac, Arizona, United States)
Goodenough Community (Seattle, Washington, United States) (Tahuya, Washington, United States) Re-Forming
Govardhan Ecovillage (Galtare, Maharashtra, India)
Govinda Gardens (Carriere, Mississippi, United States)
Green Acres Neighborhood Ecovillage (Bloomington, Indiana, United States) Forming
Green Sage Tribe (Holland, Michigan, United States) Forming
Green Valley Village (Sebastopol, California, United States) Forming
Green Village Philadelphia (GVP) (Philadelphia, Pennsylvania, United States) Forming
Greenplan (Berkeley, California, United States) Re-Forming
Greensoul (Klamath Falls, Oregon, United States) Forming
Grishino Community (Leningradskaya oblast', Russian Federation)
The Grove (Mesa, Arizona, United States) (Cocoroche Island, Belize) Forming
Gypsy Haven (Nelson, British Columbia, Canada) Forming
Hamster In The Sheets (Lake Placid, Florida, United States) Forming
Harmony Green Village (Delaware County, Pennsylvania, United States) (Delaware,

United States) Forming

HarmonyCollective (Ann Arbor, Michigan, United States) Forming

Healing Earth New Amish (Granite Shoals, Texas, United States) (near Fredericksburg, Texas, United States) Forming

Heart and Spoon Community House (Eugene, Oregon, United States)

Heartland Ecovillage (Loveland, Ohio, United States) Forming

HeartSong EcoVillage At SpringWater Community (Ravenden, Arkansas, United States) Forming

Heartspace Ecovillage (Austin, West Coast, BC Canada, Pacific NW, Rocky Mountains Area, etc., Texas, United States) Forming

Hedonisia Hawaii Sustainable Community Rainforest Retreat (Pahoa, Hawaii, Big Island, United States)

Hertha (DK-8464 Galten, Denmark) Forming

Hickory Grove (Hopkinsville, Kentucky, United States) Forming

Hidden Forest Dwellers (Saskatoon, Saskatchewan, Canada) Forming

Hilo Cohousing (Hilo, Hawaii, 96720, United States) Forming

Holy Angels Co-op and eco-village (Toledo, Ohio, United States) Forming

Home Ecovillage Project (Brighton, England, United Kingdom) Forming

Home Tree (Henniker, New Hampshire, United States) Forming

Home without Marriage and Family(1st Branch) (Anning, China) (Chuxiong, Yunnan, China) (Lincang, China)

Homeland (Thora, New South Wales, Australia)

Honua Oia'i'o Kauai (Anahola, Kauai, Hawaii, United States) Forming

Hopewell Community (Lanesville, Indiana, United States) Re-Forming

Hudson Valley Ecovillage (Hudson Valley, New York, United States) Forming

Hummingbird Community (Mora, New Mexico, United States)

I-City (Burlington, Colorado, United States) (Kanorado, Kansas, United States) (Goodland, Kansas, United States)

IICF (Ahmedabad, Gujarat, India) Forming

Imagine Seven (Progressive Communities Sharing One Property in One Community) (Olympia, Washington, United States) Forming

In Planning Ecovillage (Zuidbroek, Netherlands) Forming

InanItah (Isla de Ometepe, Nicaragua, Ometepe, Nicaragua) Forming

Indigo (Big Island of Hawaii) (South Point, Hawaii, United States) Forming

INFINITY NV EXPERIMENTAL BUILDING FARM (Battle Mountain, Nevada, United States) Forming

Inkiri of Piracanga EcoCommunity (Itacaré, Bahia, Brazil)

Inspiring Solidarity (Skattungbyn, Orsa, Sweden) Forming

Iowa City Co-Housing (Iowa City, Iowa, United States) Forming

ISLOVE = infinite star light offering visionary ecovillages (Santa Cruz, California, United States)

italyecovillage, Le Mogli farm (Pescosolido, Italy) Forming

the jaguar tribe (SAN AGUSTIN, HUILA, Colombia) (SAN AGUSTIN, HUILA, Colombia) Forming

Jamaica Plain Cohousing (Boston, Massachusetts, United States)

Jewel Creek Ecovillage (Greenwood, British Columbia, Canada) Forming

Jewish Eco Village (El Cerrito, California, United States) Forming

Juniper Hill Farm (Amherst, Massachusetts, United States)
Kailash Ecovillage (Portland, Oregon, United States)
Kakwa Ecovillage Cooperative (McBride, British Columbia, Canada) Forming
Kalikalos (Pelion, Magnesia, Greece)
KanAwen (Canet d'Adri, Girona, Spain) Forming
Kanjini Co-Op (Queensland, Australia) Forming
Kapievi (Puerto Maldonado, Madre de Dios, Peru)
Katywil Farm Community (Colrain, Massachusetts, United States) Forming
Kawai PuraPura (Albany, Auckland, New Zealand) Re-Forming
Kibeti Ecovillage (Kibeti, Congo, The Democratic Republic of The)
Kitezh Children's Eco-Village Community (249650 Baryatinsky, Russian Federation)
Kivisalon yhteisökylä (Kuopio, Finland) Forming
Kohatu Toa Eco-Village (Northland, New Zealand)
Komkelen (Buenos Aires, Capital Federal, Argentina) Forming
Kookaburra Park Eco-Village (Gin Gin, Queensland, Australia) (Bundaberg, Queensland, Australia) Forming
La Cité Écologique de Ham-Nord (Ham-Nord, Quebec, Canada)
La Ecovilla (San Mateo, Alajuela, Costa Rica) Forming
La Paz Eco Village (La Paz, Baja California Sur, Mexico) (Baja California Sur - Mexico, Mexico) Forming
Lah lah land (Tucson, Arizona, United States) Re-Forming
Lake Chapala, Mexico (Chapala, Jalisco, Mexico) Forming
Lake Claire Cohousing (Atlanta, Georgia, United States)
Lammas (Pembrokeshire, Wales, United Kingdom) Forming
Lanark Ecovillage (Lanark, Ontario, Canada) Forming
laSenda Ecovillage (San Miguel de Allende, GTO, Mexico) Forming
Lavender Falls (Fairfield Glade, Tennessee, United States) Forming
Laytonville Ecovillage (Laytonville, California, United States) Forming
LE CASE Ecovillage (San Diego, United States)
Lebensgarten Steyerberg (Steyerberg, Lower Saxony, Germany)
Lemuria Center (Centro Lemuria / Lemuria Center, Rancho Cajones, Guanajuato, Mexico) Forming
Lethbridge Eco Co-op (Lethbridge, Alberta, Canada) Forming
The Light Center (Baldwin City, Kansas, United States) Forming
Lightwork Ecovillage (Hope Point, Gambier Island, British Columbia, Canada) (Vancouver, British Columbia, Canada) Forming
Little Wonder Falls Earthship & Community Hub (Horning's Mills, Ontario, Canada)
Living Roots Ecovillage (Jasper,, Indiana, United States) Forming
Living Well Community (Franklinville, North Carolina, United States) Forming
Loh El (Ottawa, Ontario, Canada) Forming
Los Angeles Eco-Village (Los Angeles, California, United States)
Los Visionarios (Vilcabamba, Loja, Ecuador) Re-Forming
Lothlorien (Baturaden, Central Java, Indonesia)
The Love Israel Family (Kettle Falls, Washington, United States) Forming
Lovetribe (Portland, Oregon, United States) Forming
LUBINKA (Moscow region, Russia, Russian Federation) Forming
Mandala (Maryvale, Queensland, Australia)

Manitou Arbor Ecovillage (Kalamazoo, Michigan, United States) Forming
Me Lucky Farms (Blaine, Washington, United States) Forming
Meadowsong Cohousing Ecovillage Forming
Mele Nahiku (Nahiku, Hawaii, United States)
The Metahive Project (Summerland, British Columbia, Canada) Forming
metasofa artists community (Tucson, Arizona, United States) Forming
Metro Cohousing at Culver Way (St. Louis, Missouri, United States) Forming
Michigan EcoVillage (Flint, Michigan, United States) (Detroit, Michigan, United States) Forming
Miracle Springs (Nong Kai, Thailand, Thailand) Forming
Moana 'ula - Low Fat, Raw Vegan Ecovillage (Kapoho, Hawaii, United States) Forming
Moonlight Meadows (United Kingdom) Re-Forming
Mosaic Commons (Berlin, Massachusetts, United States)
Mosaic Ecovillage (Vancouver/Camas, Washington, United States) Forming
Moshav Ecovillage (United States) Forming
Mother Earth (Arizona, United States) Forming
Mount Eden Ecovillage (Washington, New Jersey, United States) Forming
Mulvey Creek Land Co-operative (Slocan, British Columbia, Canada) Re-Forming
Nahziryah Monastic Community (Saint Joe, Arkansas, United States)
Namasté Greenfire (Center Barnstead, New Hampshire, United States) Forming
Natural Fibre Enterprise Village (tbc) (Armstrong to Salmon Arm Region, British Columbia, Canada) Forming
Natural Island Dragonmill (Schweta, Saxon, Germany) Forming
Nature Island Ecovillage (Jacco Estate, Belles, Dominica) Forming
The Nature School Foundation Inc. (Greenville, New Hampshire, United States) Forming
Naturist Eco Village (south of Europe) Forming
Networking For Peace Multicultural Cohousing Resource Neighborhood (Portland, Oregon, United States) Forming
New England Farm Village Project (Maine, United States) Forming
New Talavana (Carriere, Mississippi, United States)
Newberry House (Portland, Oregon, United States)
NEXT EVOLUTION COMMUNITY (PA, OH, WVA, NC, Pennsylvania, United States) Forming
Next Step Integral (Winlaw, British Columbia, Canada)
no where ranch (Teabo, Yucatan, Mexico) Forming
NW NJ Ecovillage (Sussex County, New Jersey, United States) Forming
O.U.R ECOVILLAGE (Shawnigan Lake, British Columbia, Canada) Forming
Oasis Eco-Village (Uhland, Texas, United States) Forming
Oceanic Ecovillage (c) (Louisiana, United States) Forming
Odonata (Newburyport, Massachusetts, United States) Forming
Okie Farms Ecovillage (Oklahoma, Oklahoma, United States) Forming
Olympia Eco-Village (Olympia, Washington, United States) Forming
OMfield (Bend, Oregon, United States)
One Island Sustainable Living Center (Honaunau, Hawaii, United States)
Ontario Eco-Village (s) Project (s) (Ontario, Canada) Forming
Opihikao Ecovillage/Cohousing (pahoa, Hawaii, United States) Forming

OQuinn mountain village nomadic community (Roseville, California, United States)
Orange Twin Conservation Community (Athens, Georgia, United States) Forming
Oregon Working Group 2006 (Eastern Oregon, Oregon, United States) Forming
Otamatea Eco-Village (Kaiwaka, Northland, New Zealand)
Pagan Intentional Community (Tulsa, Oklahoma, United States) Forming
The Paradise Builders (Malibu, California, United States) (Levuka, Fiji) (Jaco, Costa
Rica) (David, Panama) (Byron Bay, Australia)
Paradise in Argentina (near Merlo, San Luis, Argentina, San Luis, Argentina) Forming
PAZ Ecovillage (Terlingua, Texas, United States) (Texas, United States) Forming
Peacefull Village (Southern Ontario near Guelph hopefully!!, Ontario, Ontario, Canada)
Forming
Penyon Bay Ecological Village (Alhoceima, Morocco) Forming
Permalogica (Tabua, Portugal) Forming
Pilot House (Cincinnati, Ohio, United States) Forming
Piñon Ecovillage (Santa Fe, New Mexico, United States) Re-Forming
Plants for a Future (Ashwater, Beaworthy, Devon, England, United Kingdom) (The
Field, Higher Penpol, Lostwithiel, Cornwall, England, United Kingdom) Forming
Plenitude Ecovillage (Madison, Wisconsin, United States) (Viroqua, Wisconsin, United
States) Forming
The Point of Infinity Retreat Center and Community (Greenfield Park, New York,
United States) Forming
Port Townsend EcoVillage (Port Townsend, Washington, United States) Forming
Pot Luck Farms (Westtown, New York, United States) Forming
Prairie Creek Settlement (Lebanon, Missouri, United States) Forming
Prairie's Edge Eco-Village (River Hills, Manitoba, Canada)
Praterra (boa vista roraima, Brazil) Forming
Primitive Self-Sufficient World-Saving Enlightencamp (oregon city, Oregon, United
States) Forming
Primitive Ways Ecovillage (California, United States) Forming
The Priorities Institute (Denver, Colorado, United States) Forming
Project Nuevo Mundo (Santiago Atitlan, Solola, Guatemala)
PROJETORGONE (Neuilly sur marne, France) Forming
Queer Ecovillage (Asheville, North Carolina, Formerly Gainseville, FL, United States)
Forming
The Rabbit's Hole Village (Southern CO. / Not purchased., Colorado, United States)
Re-Forming
Rainbow Cottage (Phoenix, Arizona, AZ, United States) Re-Forming
The Rainbow Hostel (Seeking Land, Costa Rica) Forming
Rainbow Peace Caravan (Mobil, Mexico)
The Rainbow Ranch (colorado Springs, Colorado, United States) Forming
Rainbows End (Koro Island, Fiji) Forming
Red Earth Farms (Rutledge, Missouri, United States)
Réseau des ÉcoHameaux et ÉcoVillages du Québec (Wotton, Quebec, Canada)
Forming
The Revolution Project (Chimirol de Rivas, Costa Rica) Forming
Rhiannon Community (Malchingui, Ecuador)
Riddle Farm (Marshall, North Carolina, United States) Forming

Rivendell (New South Wales, Australia) (Queensland, Australia)  Forming
Rockland EcoVillage & Cohousing (New York, United States)  Forming
Rosy Branch (Black Mountain, North Carolina, United States)
Sacred Garden Sanctuary (Douglas, Arizona, United States)  Forming
Sadhana Forest (Auroville, Tamil Nadu, India)
sailing the farm - sailing/boatbuilding coop. (Oslo, Norway, Norway)
Salon de Montclair (Salinas, California, United States)  Forming
San Francisco Backyard Neighborhood and Learning Center (San Francisco, California, United States)  Forming
San Mateo Ecovillage (San Mateo, California, United States)  Forming
Sandhill Farm (Rutledge, Missouri, United States)
Sat Yoga Eco-Village (Perez Zeledon, Costa Rica)  Forming
Savita Lodge (Uruti, Taranaki, New Zealand)  Forming
Sawyer Hill EcoVillage (Berlin, Massachusetts, United States)
The School of Love-Awareness  Forming
SE-ADK-coCamping-coHousing (Clemons, New York, United States)  Forming
SEEEDs (Gloucester, NSW, Australia)  Forming
Seven Acres Cooperative (Soquel Hills, Santa Cruz, California, United States)
Sharing Circle Community (Cincinnati, Ohio, United States)  Re-Forming
Sieben Linden Ecovillage (Beetzendorf, Germany)
Sircadia (Quillabamba, Peru)  Forming
Sirius Community (Shutesbury, Massachusetts, United States)
611 Ecovillage (Oakland, California, United States)
Skyhouse Community (Rutledge, Missouri, United States)
SoFair Farms (Fairfield, Iowa, United States)  Forming
Sólheimar (Selfoss, Iceland)
SomerVille Ecovillage (Chidlow, Western Australia, Australia)  Forming
The Source Farm Ecovillage (Jamaica)  Forming
Sovereign Christian Patriot Mission (ABOVE BAÑOS, Ecuador)  Re-Forming
The Spirit Project (New England Area)  Forming
Spiti ton Kentavron | Pelion Holistic Education Centre (Anilio - Pilion, Greece)  Forming
St Croix Valley Highlands (Prescott, Wisconsin, United States)  Forming
Stardust Center for Sustainability and Community (Gainesville Area, Florida, United States)  Forming
Stargazer EcoVillage (Greenville, South Carolina, United States)  Forming
Star*Light Retreat, South Africa {Vegetarian. clean sober smoke-free = no drugs/alcohol/smoking} (rural savannah called veld. high elevation above Malaria level. within couple hours drive of 2 major international airports of 2 of the largest cities in South Africa., South Africa) (Santa Cruz, California, United States)
Still Spirit Community Project (Floyd, Virginia, United States)
Stone's Throw Ecovillage (Viroqua, Wisconsin, Driftless area, United States) (Viroqua, Wisconsin, United States)  Forming
Stony Brook Cohousing (Jamaica Plain, Massachusetts, United States)  Forming
Suderbyn (Visby, Gotland, Sweden)  Forming
SunBird (Golden, British Columbia, Canada)  Forming
Sundog Ecovillage (Potomac, Montana, United States)

Svanholm (DK-4050 Skibby, Denmark)
Sydney Coastal Ecovillage (New South Wales, Australia) Forming
Tamarack Knoll Community (Fairbanks, Alaska, United States)
Tamera - Healing Biotope I (7630 Colos, Portugal)
Tapestry Cohousing (Waterloo Region, Ontario, Canada) Forming
Tasman Village (Nubeena, Tasmania, Australia) Forming
TBA (Gulf Islands/Discovery Islands, British Columbia, Canada) Forming
Tempelhof (Kressberg, Baden-Württemberg, Germany) (Pullach, Germany)
TerraVie (Saint- Sauveur-des-Monts, Quebec, Canada) Forming
Thomas Farm Community (Rindge, New Hampshire, United States) Forming
Three Groves Ecovillage (West Grove, Pennsylvania, United States) (West Grove, Pennsylvania, United States) Forming
Three Rivers Tribe (Pittsburgh, Pennsylvania, United States) Forming
Tir Tairngire (Corner Brook, Newfoundland And Labrador, Canada) Forming
Tir Tairngire Project (Corner Brook, Newfoundland And Labrador, Canada) Forming
Toronto Ecovillage Project (Toronto, Ontario, Canada)
Torri Superiore Ecovillage (Ventimiglia, Imperia, Italy)
Tranquility's Way (Mentone, Alabama, United States) Forming
Transcendent Seeds of Paradise (California) Forming
Tree of Life Ecovillage and Resource Center for Nonviolence (Athens, Ohio, United States) Forming
Tribe of Likatien (Füssen, Germany)
TRINITY Christian PRIORY (Spain) (Argentina) (Vera Cruz, Mexico) (Uppsala, Sweden) (Crawley, United Kingdom) (Bebra, Germany) (RENNES, France) (Dallas, Texas, United States) (Congo) (PARIS, France) (Selignac, France) (THIERS 63300, France) (Welkom, South Africa) (New Mexico and Mexico) (Wisconsin, United States) (Chicago, Illinois, United States) (Mpumalanga Province, near Badplaas, South Africa) (Newark, Notts, United Kingdom) (Venezuela) (Brazil) (Ecuador) (India)
Tuscaloosa Cooperative Association (Tuscaloosa County, Alabama, United States) Forming
Tuwa The Laughing Fish (Cabiao, Nueva Ecija, Philippines) Forming
Tweed Valley Ecovillage Project (Scotland, United Kingdom) Forming
UHURU (Rice, Washington, United States)
Umphakatsi Peace Ecovillage (Elukwatini, Steynsdorp, Mpumalanga, South Africa)
Understenhodgen (Stockholm, Sweden)
usavillages (Richland, Iowa, United States) Re-Forming
Utah Valley Commons (Provo, Utah, United States) Forming
Valdepielagos (Valdepielagos (Madrid), Spain) Forming
Valle de Sensaciones (Alpujarras, Granada, Spain)
Valles de Nanzal (Nanzal, Panama, Panama) Forming
V-Ecovillage (Gainesville, Florida, United States)
Velvet Green (Spokane, Washington, United States) Forming
Ventura Urban Homestead Cooperative (Ventura, California, United States)
Verein Keimblatt (Zurndorf, Austria) Forming
A Village Full of Coops (Brooklyn, New York, United States) Forming
The Village Project (Victoria, British Columbia, Canada) Forming
Village Terraces Cohousing Neighborhood (Black Mountain, North Carolina, United

States)
the Villages at Crest Mountain (Asheville, North Carolina, United States)
Vine & Fig Tree Farm/Community (Lanett, Alabama, United States) Re-Forming
Vital Village (Comisaría Emiliano Zapata, carretera Xul km 12, Oxkutzcab, Yucatán, Mexico) (Vital Village, Yucatan, Mexico) Forming
Vlierhof (Kleve, Millingen, Netherlands) Re-Forming
Walden Ecovillage Forming
Walden Farm (Saskatchewan, Canada) Forming
Weltsmertz Ranch Cooperative (Kyote, Texas, United States) Forming
WHISPERING ECO-WIND FARM COMMUNITY & INSTITUTE (Berwick, Maine, United States) Forming
White Buffalo Farm Ecovillage (Paonia, Colorado, United States) Re-Forming
White Hawk (Ithaca, New York, United States) (Danby, New York, United States)
Whole Village (Caledon, Ontario, Canada)
Wholistic Community Network IRC (British Columbia, Canada) Forming
Wild Cat Creek Ecovillage (Calhoun County, 6 miles east of Bluntstown on highway 20, Florida, United States) Forming
wild solutions farm Forming
Wildflower Ecovillage (Dallas east rural, Texas, United States) Forming
Wind Walker Eco Village (Spring City, Utah, United States) Forming
WinSol3 (Somerset, California, United States) Forming
Wyomanock (stephentown, New York, United States) Forming
YAN (Copenhagen, Denmark) Re-Forming
Yarrow Ecovillage (Chilliwack, British Columbia, Canada)
yewlandia (West Kootenays, British Columbia, Canada) Forming
Yoga Eco-Village (Bali, Indonesia) (Salt Spring Island, British Columbia, Canada) Forming
The Yoga Farm (Punta Banco, Costa Rica)
ZEGG - Centre for Experimental Cultural Design (Belzig, Germany) (Bad Belzig)

# Further Reading

"Planning for a Post-Oil Economy" by Peter Goodchild.

Ashworth, Suzanne. Seed to Seed. Decorah, Iowa: Seed Saver, 1991.

Bagdikian, Ben H. The Media Monopoly. 6th ed. Boston: Beacon, 2000.

Bailey, L.H. The Principles of Vegetable-Gardening. New York: Macmillan, 1921.

Blainey, Geoffrey. Triumph of the Nomads: A History of Aboriginal Australia. Woodstock, New York: Overlook, 1976.

Bradley, Fern Marshall, and Barbara W. Ellis, eds. Rodale's All-New Encyclopedia of Organic Gardening. Emmaus, Pennyslvania: Rodale, 1992.

Broadfoot, Barry. Ten Lost Years 1929-1939: Memories of Canadians Who Survived the Depression. Toronto: Doubleday, 1973

Brown, Lauren. Grasses. Boston: Houghton Mifflin, 1979.

Bubel, Mike and Nancy. Root Cellaring. Pownal, Vermont: Storey, 1991.

Campbell, Colin J. The Coming Oil Crisis. Brentwood, Essex: Multi-Science, 1997.

Carter, Vernon Gill, and Tom Dale. Topsoil and Civilization. Rev. ed. Norman, Oklahoma: U of Oklahoma P, 1974.

Catton, William R., Jr. Overshoot: The Ecological Basis of Revolutionary Change. Champaign, Illinois: U of Illinois P, 1980.

Davis, Adelle. Let's Eat Right to Keep Fit. Rev. ed. New York: Harcourt Brace Jovanovich, 1970.

Deffeyes, Kenneth S. Hubbert's Peak: The Impending World Oil Shortage. Princeton: Princeton UP, 2001.

Ellis, Barbara W., and Fern Marshall Bradley, eds. The Organic Gardener's Handbook of Natural Insect and Disease Control. Emmaus, Pennyslvania: Rodale, 1992.

Emery, Carla A. The Encyclopedia of Country Living. 9th ed. Seattle, Washington: Sasquash, 1994.

Faust, Joan Lee. The New York Times Book of Vegetable Gardening. New York: Times, 1975.

Gardner, Sandra. Street Gangs in America. New York: Franklin Watts, 1992.

Greer - 'Facing the New Dark Age'

Gever, John, et al. Beyond Oil: The Threat to Food and Fuel in the Coming Decades. Cambridge, Massachusetts: Ballinger, 1986.

Gibbons, Euell. Stalking the Wild Asparagus. New York: David McKay, 1962.

Goodchild, Peter. Survival Skills of the North American Indians. 2nd ed. Chicago Review Press, 1999.

Gowdy, John, ed. Limited Wants, Unlimited Means: A Reader on Hunter-Gatherer Economics and the Environment. Washington, D.C.: Island, 1998.

Greenwood, Pippa. Pests and Diseases. New York: Dorling Kindersley, 2000.

Guillet, Edwin C. The Pioneer Farmer and Backwoodsman. Toronto: Ontario, 1963.

Hopkins, Donald P. Chemicals, Humus, and the Soil. Brooklyn, NY: Chemical Publishing, 1948.

Jacob, Jeffrey. New Pioneers. University Park, Pennsylvania: Pennsylvania University Press, 1997.

Kaplan, Robert D. The Ends of the Earth: From Togo to Turkmenistan, from Iran to Cambodia - A Journey to the Frontiers of Anarchy. New York: Random, 1996.

King, F.H. Farmers of Forty Centuries, or, Permanent Agriculture in China, Korea and Japan. 1911. Emmaus, Pennsylvania: Organic Gardening, n.d.

Klare, Michael T. Resource Wars: The New Landscape of Global Conflict. New York: Henry Holt, 2001.

Langer, Richard W. Grow It! New York: Saturday Review, 1972.

Lappé, Frances Moore. Diet for a Small Planet. New York: Ballantine, 1971.

Logsdon, Gene. Homesteading. Emmaus, Pennyslvania: Rodale, 1973.

——. Small-Scale Grain Raising. Emmaus, Pennyslvania: Rodale, 1977.

Mack, Norman, ed. Back to Basics. Montreal: Reader's Digest, 1981.

Meadows, Donella H. et al. The Limits to Growth: a Report for the Club of Rome's Project on the Predicament of Mankind. 2nd ed. New York: Universe, 1982.

Nearing, Helen and Scott. Living the Good Life. New York: Schocken, 1982.

Niethammer, Carolyn. American Indian Food and Lore. New York: Macmillan, 1974.

Pimentel, David, and Carl W. Hall, eds. Food and Energy Resources. Orlando: Academic, 1984.

Rifkin, Jeremy. The End of Work: The Decline of the Global Labor Force and the Dawn of the Post-Market Era. New York: Tarcher/Putnam, 1995.

Scher, Les. Finding and Buying Your Place in the Country. 4th ed. New York: Collier, 1996.

Schumacher, E.F. Small Is Beautiful: Economics as if People Mattered. New York: Harper & Row, 1989.

Seymour, John. The Guide to Self-Sufficiency. New York: Popular Mechanics, 1976.

Solomon, Steve. Water-Wise Vegetables. Seattle: Sasquatch, 1993.

Soros, George. The Crisis of Global Capitalism. New York: PublicAffairs, 1998.

Tresemer, David. The Scythe Book. Brattleboro, Vermont: Hand & Foot, 1981.

Vivian, John. The Manual of Practical Homesteading. Emmaus, Pennsylvania: Rodale, 1975.

Weatherwax, Paul. Indian Corn in Old America. New York: Macmillan, 1954.

Widtsoe, John A. Dry-Farming. New York: Macmillan, 1920.

W.C. Clark - "The real reasons for the upcoming war with Iraq: a macroeconomic and geostrategic analysis of the unspoken truth"

## THE AUTHOR

Dr. Sahadeva dasa (Sanjay Shah) is a monk in vaisnava tradition. Coming from a prominent family of Rajasthan, India, he graduated in commerce from St.Xaviers College, Kolkata and then went on to complete his CA (Chartered Accountancy) and ICWA (Cost and works Accountancy) with national ranks. Later he received his doctorate.

For close to last two decades, he is leading a monk's life and he has made serving God and humanity as his life's mission.

He has been serving as the president of ISKCON Secunderabad center since last twenty years.

His areas of work include research in Vedic and contemporary thought, Corporate and educational training, social work and counselling, travelling in India and aborad, writing books and of course, practicing spiritual life and spreading awareness about the same.

He is also an accomplished musician, composer, singer, instruments player and sound engineer. He has more than a dozen albums to his credit so far. (SoulMelodies.com) His varied interests include alternative holistic living, Vedic studies, social criticism, environment, linguistics, history, art & crafts, nature studies, web technologies etc.

He is the author of many internationally acclaimed books.

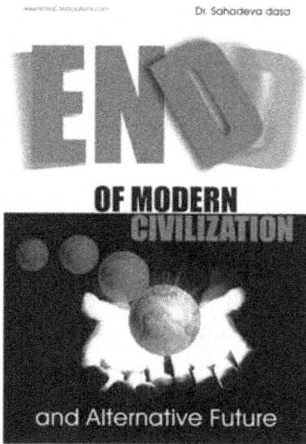

This book by Dr Sahadeva dasa is an authoritative work in civilizational studies as it relates to our future. Dr. dasa studied human civilizations of last 5000 years and the reasons these civilizations went into oblivion. Each of these civilizations collapsed due to presence of one or two factors like neglect of soil, moral degradation, leadership crisis etc. But in our present civilization, all the factors that brought down all the these civilizations are operational with many more additional ones. Then the book goes on to chalk out the alternative future for mankind.     Pages-440, www.WorldCrisisSolutions.com
For a copy, write to: soulscienceuniversity@gmail.com

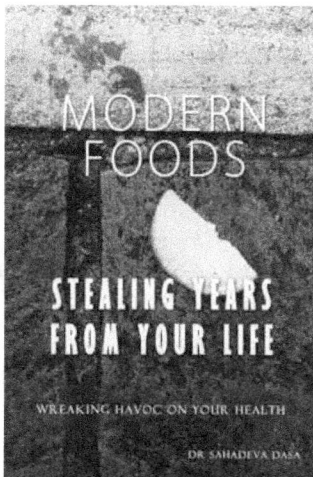

Food is our common ground, a universal experience. But there is trouble with our food. Traditional societies had good food but we just have good table manners. A disease tsunami is sweeping the world. Humanity is dying out. This is the result of our deep ignorance about our food. If you don't have good health, the other things like food, housing, transportation, education and recreation don't mean much. This books lists out major killer foods of our industrial civilization and how to escape them.     Pages 276, www.FoodcrisisSolutions.com
For a copy, write to: soulscienceuniversity@gmail.com

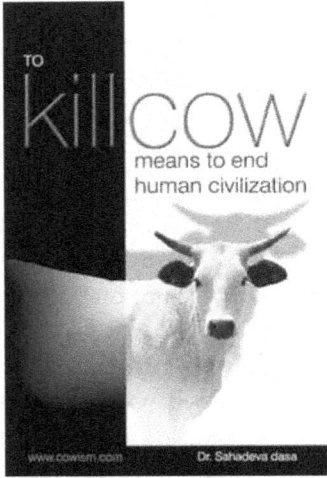

This landmark book on cow protection delineates various aspects of cow sciences as presented by the timeless voice of an old civilization, Vedas. This book goes on to prove that the cow will be the making or breaking point for humanity, however strange it may sound. Science of cow protection needs to be researched further and more attention needs to be given to this area. Most of the challenges staring in the face of mankind can be traced to our neglect in this area.

Pages-136, www.cowism.com

For a copy, write to: soulscienceuniversity@gmail.com

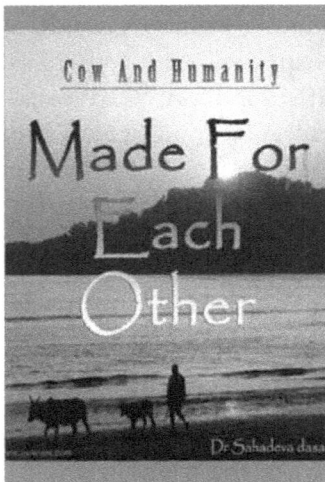

This book discusses the vital role of cows in peace and progress of human society. Among other things, it also addresses the modern ecological concerns. It emphasizes the point that 'eCOWlogy' is the original God made ecology. For all the challenges facing mankind today, mother cow stands out as the single answer.

Living with cow is living on nature's income instead of squandering her capital. In the universal scheme of creation, fate of humans has been attached to that cows, to an absolute and overwhelming degree.

Pages-144, www.cowism.com

For a copy, write to: soulscienceuniversity@gmail.com

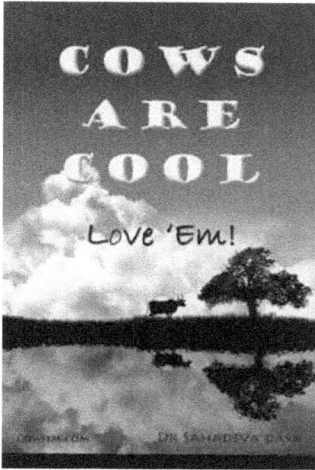

This book deals with the internal lives of the cows and contains true stories from around the world. Cow is a very sober animal and does not wag its tail as often as a dog. This does not mean dog is good and cow is food. All animals including the dog should be shown love and care. But cow especially has a serious significance for human existence in this world. Talk about cows' feelings is often brushed off as fluffy and sentimental but this book proves it otherwise.

Pages 136, www.cowism.com

For a copy, write to: soulscienceuniversity@gmail.com

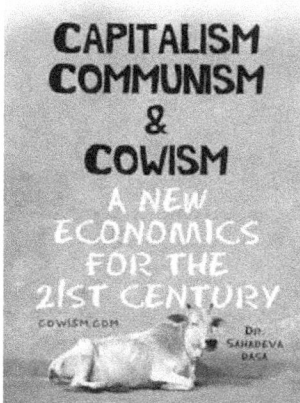

If humanity and the planet have to survive, we have to replace our present day economic model. It's a fossil fuel based, car-centred, energy inefficient model and promotes over exploitation of natural resources, encourages a throwaway society, creates social injustice and is not viable any longer.

This book presents an alternative economic system for the 21st Century. This is an economics which works for the people and the Planet.

Pages 136, www.cowism.com
For a copy, write to: soulscienceuniversity@gmail.com

www.ingramcontent.com/pod-product-compliance
Lightning Source LLC
Chambersburg PA
CBHW060004210326
41520CB00009B/817